TIME SUTRA

TIME SUTRA

Solving Time's Mystery

J. F. Johnson

Climax Canyon Press
P. O. Box 333
Raton, New Mexico 87740
jfjohnson@q.com

A brief survey and criticism of the philosophical
and scientific literature on time leading to a
deeper understanding of time and memory.

Johnson, J. F.
Time Sutra : Solving Time's Mystery
Includes glossary, bibliographic reference and index.
ISBN: 9780615836591 (paper).
1. Time (philosophy). 2. Memory.
3. Cognitive Sciences. 4. Buddhism.

Climax Canyon Press
P. O. Box 333
Raton, New Mexico 87740

Cover photo by author: detail of
1910, eight day chronometer,
Waltham Watch Company.

CONTENTS

PREFACE

The aim of philosophy? ... To show the fly the way out of the fly bottle.
—Ludwig Wittgenstein[1]

One evening in the late summer of 1984, I was sifting through an area of inquiry I had visited so many times in the past, when I saw something I hadn't previously noticed—and everything changed. Over the many months that followed, again and again I would return to this 'meditation' and slowly this practice stabilized. I began to take notes that evolved into a nightly journal. I found myself always wanting more time to engage this 'meditation'.

I had a decent day job, but it made high demands on my time, so for fifteen years I studied in the evenings, which meant fewer and fewer social contacts. I saved my money, eventually quit work, and moved to a small town in northern New Mexico. I now owned all of my time. I bought an old house and made it into a comfortable retreat. For more than fourteen years I have had the time to sit, read, and record my thoughts. Now I am prepared to say what it was I first encountered on a late summer evening so many years ago.

Early on I understood the entanglement between time and memory, and over the past quarter century I have worked out the details. For the most part, the details are learning how to speak of something that is nearly unreachable with language. Time, and this is a book about time, has its own particularly slippery character. T. S. Eliot wrote beautifully about time[2] and about how expressing time stretches language to its limit: "Words strain,/Crack and sometimes break, under the burden,/Under the tension, slip, slide, perish,/Decay with imprecision, will not stay in place,/Will not stay still."[3]   The subject of *Time Sutra* is reality, but it is a story told in words, and words are never the exact thing to which they refer; words are never reality. Language is always a self-referential system that struggles to express reality. Language utilizes simile, allegory, myth, allusion and metaphor to say that which cannot be said. I have learned the language of the debate from reading a wide spectrum of the literature on time and memory. The debate ranges from spacetime physics, psychology, cybernetics, neuroscience and to philosophy (both Eastern and Western). The following lays out the reason for the story of *Time Sutra*, the structure of the telling, and the importance of what is being told.

The thesis I am presenting is that our sense of memory is distorted. As such, this distorts our reality. The distortion is not something inherent in the sense of memory, rather, the problem is how we think about memory. The mistake lies at the very core of how we see the world. The way we habitually think creates a sense of familiarity that blinds us to the mistake. We suppose memory is something that it isn't, and with this ingrained supposition in mind, we learn to see memory through the doctrine of our human culture. *Time Sutra* is a meditation aimed at correcting this error that lies at the heart of our being. This is where the term *Sutra* comes in —*sutra* is a Sanskrit term that literally means thread. In Eastern philosophy the text of a *sutra* was used as a mnemonic device that would prompt a remembrance of the 'thread of the argument' which a particular, often lengthy, religious text was trying to convey. *Time Sutra* is intended as a 'thread of reason', which, when followed, will lead to understanding the nature of time and memory. This is a meditation that focuses on unraveling the time-memory-self nexus, and when this nexus is understood, it brings to mind something very important that we have forgotten.[4]

I argue time is an illusion, because there is no empirical evidence for time. There is no demonstrable 'arrow of time' in all of physics, and without the arrow of time, the concept of time is doomed. In fact, the majority of present-day theoretical physicists do not believe time to be a fundamental aspect of reality.[5] Implicit in the theory of modern cosmology is the 'presupposition of the *past*,' and in recent years, physicists have come to grips with this problem which has been termed *The Past Hypothesis*.[6] However, philosophers for some time have recognized a far more insidious 'presupposition of the *past*,' sometimes called *The Representative Theory of Memory*, whereby we presuppose the present experience of memory represents something that is past.[7] I will remind the reader that the past is only hypothetical by frequently invoking the suppositional phrase *memory is past*—employing this phrase as a mnemonic tool will 'bring to mind' how we have been taught to think about time and the world.

I then turn to the findings of the cognitive sciences. I draw on evidence from neuroscience, cybernetics and artificial intelligence, psychology, philosophy of mind, and linguistics, to show there is no evidence of a *self-ego*, and that the self-ego we imagine ourselves to be is a mental construction formed out of a sense of time and memory. Eastern philosophy makes a major contribution to the development of the overall thesis, however this is not discussed until the later chapters. Eastern thinking is examined because it has often focused on *memory*, *time*, and *sixth sense*, which are exactly the key elements central to the narrative of *Time Sutra*.

The philosophical method of *Time Sutra* is a radical process, but as far as possible it is empirical, and throughout it is thoroughly rational. It is radical because the time-self-memory nexus is carefully dissected and when the dissection is finished all that remains is primordial memory. Our world becomes timeless and selfless. The distortion of memory warps our reality; it is our common belief, *memory-is-past,* which causes the disenchantment[8] of the world. Time and self-ego are mental constructs, only memory is real, and memory is neither time nor self-ego.

This 'meditation' is a way of recovering memory—it is a way of 'remembering.' This is not 'remembering' in the sense of recovering something that has been forgotten. This is the recovery of the sense of memory itself, which leads to the re-enchantment of the world. This is recovering memory from the distortion that is preventing the natural experience of the sixth sense.

The argument of *Time Sutra* may seem complex, but the complexity comes from the nest of suppositions we have gathered around us. Reality is the ultimate philosophical simplicity—nothing is presumed, so experience (phenomena) is the totality of what is. This is not a radical conclusion—we will always arrive here if we employ reason. Many great philosophers have stated that all knowledge is ultimately reducible to perception alone; this includes not only the five bodily senses, but also mental perception.

As the illusion captivating our imagination dissipates, it becomes clear that *reality* is a *unity*. But to experience reality, the self-ego must be eliminated in order to heal the split that separates us from our own nature. It is the self-ego's misunderstanding about memory and time that splits the unity into self and other. The later chapters of *Time Sutra* will continue to accumulate further evidence that supports the thesis that *time* and *ego* grow out of a distortion of memory. But more importantly these later chapters are a meditation on how to extract ourselves from the trap we are in. How can we come to think about and fully understand this problem? How does the fly think its way out of the fly-bottle? Wittgenstein's fly-bottle metaphor is pointing out that the fly (by natural habit) is making assumptions that will kill it. The question becomes, how can the fly *think* in such a way it understands the nature of the fly-bottle, and so escape death? Humans, likewise, are trapped in a very similar and destructive way of thought. We are bottled-up in our habitual way of thinking. *Time Sutra,* is a meditation on changing our mind so it is more in conformity with reality. We are already in the Garden—we just need to wake up.

Time Sutra is organized into seven parts which, taken together, provide an overall structure proceeding from the beginning of our understanding time to resolving the enigma of time. Each chapter is a separate essay, but each contributes a further piece to the explanation. First up is Part I: *Beginnings,* which lays the ground for all later discussions and then initiates the argument that the answers necessary to understanding time are to be found at the *beginning.* Part II: *Anamnesis* (a penetrating inquiry into memory) confirms the time-self-memory connection, then explores this relationship between time, self, and memory, and realizes the nature of memory is found at the beginning. Part III, *Finding Time*, delves into various modern physical constructions of time. *Thinking About Time,* Part IV, presents three examples of temporal thinking as seen through the arguments of three noted philosophers. Part V: *The Path,* contains three essays, each a different meditation on time and self. Then, in Part VI: *Consequences of Time* is considered from two perspectives. Two final essays in Part VII: *Resolving Time,* gesture toward the path that leads back to the Garden.

Time Sutra would not have been written had it not been for what I have come to think of as The Writers Group. I was invited to join this small informal group of four and sometimes five. Through them I discovered how to write *Time Sutra*. I offer thanks to Pat, Richard, Sandy, and Steve for sharing your editing and writing skills. And to Kester, my friend in Australia, who provided a critical late edit of the MS. And thanks to Elaine, my longtime best friend, for your support. I simply would not have written *Time Sutra* without the help and companionship of these people.

Part I

Beginnings

Chapter One

Introduction To Time

Meditation on time is recommended as a method
for getting beyond time to the timeless reality.
—T.M.P. Mahadevan[1]

The Nature of Time: Time is the greatest of enigmas. Philosophers and physicists, since antiquity, have puzzled over the problem. Understanding the nature of time is the central problem of philosophy, and providing a physical explanation of time is the central challenge of modern physics. Although much has been written about time over the past two and a half millennia, there is no consensus as to what constitutes the nature of time. Nowhere is there a clearly stated explanation of time. *Time Sutra* is written with the intent of demonstrating a clear understanding of the phenomena of time.

The present chapter is meant to convey the flavor and variety of language employed when discussing time—it presents, in the briefest way, a sampling of the diversity of thoughts and ideas that will become important later in the text.

There is no short and simple way to characterize the debate about temporality. In fact, even a comprehensive introduction cannot possibly cover the breadth and depth of the discussion that has continued in both the East and the West for at least 2,500 years. However, an examination of several distinct recurring issues within the broader debate will provide a sense of the range of past deliberations. For the purposes of *Time Sutra*, only a cursory survey of the literature is necessary. However, from this review, the important arguments will be made apparent. Five temporal topics, having a direct bearing, will be included in this brief introductory survey. Rather than try to encapsulate each topic with a synoptic description, the five topics will be explored via comments from various thinkers who have given these issues deep consideration.

The five topics central to understanding the *nature of time* are: Importance of Time; Physics of Time; Inscrutable Time; Wisdom and Time; Unreality of Time. These topics exemplify five characteristics of the temporal debate that are of the greatest concern to *Time Sutra*. The five in no way exhaust the range of debates about time, but they include the points needed for an understanding of time. Because each of these subjects requires a more thorough investigation, a detailed analysis will be given in later chapters of the text. Meanwhile, the following montage of quotes will suffice, and the later, more fully detailed argument will bring to light the significance of these particular aspects of time.

Importance of Time: Insight into time is considered the key to understanding many scientific and philosophical problems, but even more important, it is impossible to know reality without knowing the nature of time. The following are words from a few of those for whom time was the ultimate puzzle.

It was French philosopher Henri Luis Bergson (1859–1941), "who held that understanding the nature of time is the key to the main problems of philosophy."[2] and French philosopher Jean-François Lyotard (1924–1998) who said, "It will be noted that many paradoxes belong more or less closely to the problematic of time."[3] Quoting American historian and social philosopher Lewis Mumford (1895–1990), "The clock, not the steam-engine, is the key-machine of the modern industrial age ... "[4] British philosopher David Wood (1946–) said, "There are philosophical topics other than time. Some of them engage our attention for extended periods. But sooner or later each is sucked down into the vortex of that one problem that is truly permanent: time."[5] Australian-born British philosopher Samuel Alexander

4

(1859–1938) said, "It is not, I believe, too much to say that all the vital problems of philosophy depend for their solution on the solution of the problem what Space and Time are and more particularly how they are related to each other."[6] Jean-François Lyotard also stated, "The avant-garde task remains that of undoing the presumption of the mind with respect to time."[7] And a final pronouncement on the importance of time by British psychiatrist Maurice Nicoll (1884–1953): "It has been said repeatedly that we cannot understand anything rightly unless we overcome the 'illusion of time'."[8]

Physics of Time: There is an old concept that entails thinking time has a direction—that time moves from the past through the present and into the future. This direction of the flow of time has commonly come to be referred to as the "arrow of time."[9] The modern cosmological model as expressed in the Big Bang Theory holds that the birth of the universe occurred about 13.7 billion years ago. It is thought the arrow of time extends from that distant past up to now and points into the future. However, all laws of physics are time symmetrical, so there is no implied arrow. The arrow of time in modern physics rests mainly on the assumption that the concept of *entropy*[10] provides the evidence for the direction of time. The problems associated with the "entropy assumption" will be examined in detail in Chapter 4. Following are a few of the many voices that have recognized both the importance and the difficulty of justifying an arrow of time.

Australian philosopher Huw Price (1953–) stated that "The arrow of time is one of the big unclaimed prizes of modern physics."[11] And British physicist Julian Barbour (1937–) wrote, "The arrow of time ... Since Boltzmann's age, it has towered there, an unscaled Everest. It can be described but not yet explained."[12] Columbia University theoretical physicist Brian Greene (1963–) said, " ... space and time have been the sparkling jewels of physics. They are at once familiar and mystifying; fully understanding space and time has become physics' most daunting challenge and sought-after prize."[13] Research scientist Peter Coveney (1960–) and journalist Roger Highfield (1958–) write that the arrow of time is critical to modern physics because "The arrow of time is essential for preserving the integrity of science."[14] and philosopher of physics Henry Mehlberg (1904–1979) said, "it comes as no surprise that the assumption of an arrowless, isotropic time is often viewed as a philosophical disaster..."[15]. Author and physician Larry Dossey (1940–) put it succinctly, "The view of time from modern physics tells us our ordinary notions of time are wrong."[16]

Inscrutable Time: We are all familiar with time—it is the way we think about the world. But to think about time itself brings instant consternation. Everyone can clearly see the consequences of time, yet time itself seems almost invisible. What then, is time? Our clocks are everywhere; aren't they proof of time? We think time is what the clocks measure. But we have no real sense of what is being measured. American psychologist Robert Ornstein (1942–) said, "For an analysis of the experience of time, one can point neither to an organ of perception, like the eye, nor a physical continuum, like the wave length of light for study by objective means. There is no immediate point of departure for a scientific analysis of time experience."[17] What, if anything, do the hands on the face of a clock tell us about reality? Trying to think about time is in the realm of the ungraspable. Perhaps this is a clue that time is not what it seems. Time is definitely *something*—but what is it? Again, a sampling of thoughts from the literature strongly suggests a deep mystery at the heart of the nature of time.

Philosopher Christoph Hoerl and psychologist Teresa McCormack write, " ... the ability to represent time is one of the most fundamental and least understood abilities constitutive of both consciousness and self consciousness."[18] Indian writer and philosopher Jiddu Krishnamurti (1895–1986) asked, "... is it possible for the mind to live completely, totally, in the present? It is only such a mind that has no fear. But to understand this, you have to understand the structure of thought, memory and time."[19] Argentine writer, essayist and poet Jorges Luis Borges (1899–1986) writes, "Over and over I told myself that time — that infinite web of yesterday, today, the future, forever, never — is the only true enigma."[20] Toward the end of his masterpiece, *Being and Time,* German philosopher Martin Heidegger (1889–1976) wrote, "... in the ordinary conception of time something has been covered up ..."[21]

English philosopher-mathematician Alfred North Whitehead (1861–1947) said, "It is impossible to meditate on time and the mystery of the creative passage of nature without an overwhelming emotion at the limitations of human intelligence."[22] Philosopher John McDermott wrote, "Nothing is better known, paradoxically, than the utter mystery of the meaning of time ..."[23] German philosopher and psychologist Franz Brentano (1838–1917) asked, "What is time? There is no other name that is more familiar to us, and none that is at the same time so obscure."[24] When measuring time, Whitehead commented, "Our difficulties only begin when we ask what it is that is measured."[25] The physicist Brian Greene remarks, "Time is among the most familiar yet least understood concepts that humanity has ever encountered."[26] American philosopher Jacob Needleman (1934–) asks, "Time? Surely, time is not what we think it is. We are wrong about so many lesser things; how could we imagine we understand the greatest of all mysteries, time?"[27] Needleman also said, "Time is stronger than any exercise, than any words. Time remains the great incomprehensible problem of life and thought."[28]

Wisdom and Time: The Wisdom traditions (Buddhist, Taoist, Vedic, Eleatic, etc.) are closely associated with the analysis of time. There have long been claims that wisdom and salvation are intimately involved with the understanding of time. The variety of voices that have expressed this sentiment, across so many times and cultures, is surprising. How is it that *time* and *wisdom* are related? The following array of quotes are no answer to the question, but they do demonstrate the broad acknowledgment of the prevalent time-wisdom relationship.

Professor of East Asian philosophy Steven Heine remarks, "The attainment of Buddhahood is obstructed by an inauthentic and inadequate understanding of time ..."[29] British astrophysicist Arthur Eddington (1882–1944) states, "In any attempt to bridge the domains of experience belonging to the spiritual and the physical sides of our nature, Time occupies the key position."[30] Renzai Zen Master Daio (1235–1309) said, "If you want to know the meaning of enlightened nature, you must watch the causal relations of time."[31] Dutch philosopher Baruch Spinoza (1632–1677) felt that, "To pass from the divine to the human perspective is to pass from the timeless to time, and conversely."[32]

Romanian historian and philosopher Mircea Eliade (1907–1986) stated that "Deliverance from this world and attainment of salvation are tantamount to a deliverance from cosmic Time."[33] Samuel Alexander said that "To realize the importance of Time as such is the gate of wisdom."[34] American-British poet and playwright T. S. Eliot (1888–1965) wrote in verse: "Men's curiosity searches past and future/and clings to that dimension. But to apprehend / The point of intersection of the timeless / With time, is an occupation for the saint."[35]

Irish novelist, playwright and poet Samuel Beckett (1906–1989) calls it "... that double-headed monster of damnation and salvation—Time."[36] And Beckett also remarks that "Memory and Habit are attributes of the Time cancer."[37] Klaus Klostermaier writes, "The ancient Indian text Maitri Upanisad [2nd- 3rd centuries BCE] contains the famous passage, 'Time cooks all things in the great self. He who knows in what time is cooked is the knower of the Veda'(VI,15)."[38] Indian Vakyap Bhartirhari [5th century BCE] said "In a state of ignorance it [time] is the first thing to manifest itself, but in the state of wisdom it disappears."[39]

8

Spanish scholar of comparative religion Raimundo Panikkar (1918–2010) remarks that "The sage is one who has already crossed to the other shore of the river of time."[40] Jacob Needleman makes the point "... one cannot go far with great metaphysical questions, such as the question of time, without the hard work of confronting one's own inner emptiness."[41] Practitioner of Tibetan Buddhism Charles Genoud offers a short verse, "Time veils our homeland—disguising it—as a foreign place."[42] German theologian, philosopher and mystic Meister Eckhart (c. 1260 - c. 1327) said, "Time is what keeps the light from reaching us. There is no greater obstacle to God than time."[43] Soto Zen roshi Dainin Katagiri (1928–1990) said, "No matter how long you try to make your life meaningful, you cannot find a way to do it unless you can face the original nature of time,"[44] and he also said, "At the bottom of suffering is time."[45] The great Zen master Huang Po (d. 850 CE) announces that, "Beginningless time and the present moment are the same. There is no this and no that. To understand this truth is called complete and unexcelled enlightenment."[46]

Unreality of Time: The first systematic philosopher in the Western tradition, Parmenides, denied the reality of time. During that same era in the East, the Buddha established a tradition that has denied the reality of time for the last 2,500 years. Time is such an apparent aspect of our experience that it is difficult to even consider how it might be different. Yet, there is a broad spectrum of thinkers that have doubted the existence of time.

The great 2nd century Buddhist philosopher Nagarjuna wrote, "this world of illusion ... has a nature beyond time".[47] English playwright and novelist Somerset Maugham (1874–1965) wrote, "Reality ... It is eternal because its completeness and perfection are unrelated to time."[48] German philosopher Georg Wilhelm Friedrich Hegel (1770–1831) believed that "... ultimate reality is timeless, and time is merely an illusion generated by our inability to see the whole."[49] German philosopher Friedrich Nietzsche (1844–1900), influenced by Hegel, writes, "It would even seem as if a whole diversity of things were really all of a piece, and that time is only a cloud which makes it hard for our eyes to perceive the oneness of them."[50] German-born Canadian resident and popular author Eckhart Tolle (1948–) asks, "Can there be any doubt that psychological time is a serious and dangerous mental illness?"[51] Philosopher H. S. Prasad states, "the root cause of the view of time as a reality lies in our ignorance about the nature of reality."[52] Philosopher David Wood said, "It is perhaps a common and recurrent claim among philosophers and others who think about time that there may be something wrong with 'our ordinary concept of time'."[53] French philosopher Jacques Derrida (1930–2004) offers this: "Time ... gives nothing to see. It is at the very least the element of invisibility itself."[54] Recently, philosopher of science Craig Callender (1968–) stated that "many in theoretical physics have come to believe that time does not even exist."[55] Lastly, Albert Einstein (1879–1955) famously called time a "persistent illusion."[56]

Those who examine it closely come to think time is not real, but it seems impossible to think that it is not a sense of *something*. Time *is* something, but only a rare few have ever known the truth about time. However, there have been many who have speculated on what it is that lies behind our *sense* that seems like time. More than 120 years ago, philosopher Herbert Nichols (1852–1936) expressed his exasperation at the proliferation of speculative ideas regarding the nature of time:

"Casting an eye backwards we can but be struck by the wide variety of explanations offered for the time mystery. Time has been called an act of mind, of reason, of perception, of intuition, of sense, of memory, of will, of all possible compounds and compositions to be made up of them. It has been deemed a general sense accompanying all mental content in a manner similar to that conceived of pain and pleasure. It has been assigned a separate, special, disparate sense, to nigh a dozen kinds of 'feeling', some

10

familiar, some strangely invented for the difficulty. It has been explained by 'relations', by 'earmarks', by 'signs', by 'remnants', by 'struggles', and by 'strifes', by 'luminous trains', by 'blocks of specious-present', by 'apperception'. It has been declared *a priori*, innate, intuitive, empirical, mechanical. It has been deduced from within and without, from heaven, and from earth, and from several things difficult to imagine as of either."[57]

Nichols' informative litany of speculation from 1891 misses all the abundant speculation that has occurred since that time. Certainly there has been plenty of opportunity for speculation—Samuel L. Macey's *Time: A Bibliographic Guide* estimates that in the first nine decades of the 20th century "... about 95,000 books have been published on time-related subjects."[58] Any author commenting on time-related subjects who doesn't know the true nature of time is, to a degree, speculating.

What has been said so far should demonstrate that if you don't know what *time* is, then you can't possibly know what *reality* is. All the above citations form a diversity of comments that can only come into focus when the origin of time is finally understood, but to get to this understanding we must first turn back to the beginning. There is an error at the beginning of time, and this error completely alters our world. Uncovering this error can only be approached on a path that goes to the source of time.

Chapter Two

The Fall

Wisdom is 'the Beginning' ...
—St. Augustine, 354–430 CE[1]

The allegory of the Fall is arguably the earliest and most familiar story in the Western Tradition that attempts to explain the origin of our human circumstances[2]. However, throughout the world and throughout time, cultures tell a similar story recognizable as a story of the fall and characterize it as a schism, loss, or even catastrophe. Most cultures and religions have a Creation Myth, and accompanying their narrative of the creation is a story of the fall. The tale is intended to explain the deep feeling of discontent that infects most of humankind. We feel an unease, as if we have been separated from our own nature. This affliction has come to be known as the "human condition"—a condition beset with suffering, discontent, and a complete lack of meaning, into which we are born and which only comes to an end at the time of our death. This dissatisfaction is the primary driving force behind all religion and philosophy.

The story of the fall has common attributes found across many cultures. A closer examination of these shared attributes, *time* and *timelessness, memory* and *forgetting*, will provide an insight into the loss of consciousness of our own nature and insight into why our temporal existence seems estranged from nature. The schism is often recognized as a "fall into time."[3] The schism occurs at the beginning, and prior to the schism is a timeless realm variously referred to as Sacred Time or Great Time.[4] The best clue that appears to distinguish the timeless from the temporal is something about memory—something

13

that changes as a result of the fall. Fallen *being* has a vague but pervasive feeling that something has been forgotten. The advent of the fall forms a nexus of closely tied impressions including the human condition, along with the concepts *time* and *timeless*ness, *memory* and *forgetting*. These terms describe aspects of the phenomena of the fall, but their relationship is left unanswered.

The state of unease and distress that is a consequence of the fall serves as the impetus for wanting to understand the fall, but does not in itself provide an insight into the cause of the problem. The biblical version from Genesis tells that man has offended God and God is punishing man by expelling him from his natural state. God has evicted man from Paradise. The story of "eating the fruit of the tree of knowledge" which results in "knowledge of good and evil" seems like it might be a promising allegory that explains the cause of the fall, but not without knowing a lot more about 'time and timelessness, memory and forgetting'. Author Loren Eisley writes, "The story of Eden is a greater allegory than man has ever guessed."[5]

The story of Genesis is thought to have been written between 1000 - 560 BCE[6]. The later date is consistent with the Axial Age proposed by Karl Popper, and about which Richard Geldard says "... ancient teachings from many cultures ... around 500 BCE, begin with the understanding that human beings are born possessing the memory of the divine ground to which they promise one day to return, needing only to remember what has been forgotten in the process of coming into earthly existence."[7] Karen Armstrong in *The Great Transformation* tells how "Socrates led them to discover an authentic knowledge within, which had been there all along. When this finally came to light, it felt like the recollection of an insight that had been forgotten."[8] Armstrong bracketed the Axial Age from 900 - 200 BCE[9] which is inclusive of the lives of the Buddha, Lao Tzu, Confucius, Heraclitus, Parmenides, and Socrates, and is inclusive of the ancient foundations of all modern philosophy and religion. Parmenides in particular focuses on

14

memory—the Goddess who narrates Parmenides' poem *On Nature* is the Goddess of Memory. Heidegger delved deeply into the work of Parmenides, convinced there was something back at the beginning we no longer remembered. In his inimitable style he writes, "In forgetting not only does something slip from us, but the forgetting slips into a concealment of such a kind that we ourselves fall into concealedness precisely in our relation to the forgotten."[10]

Many primitive cultures interpret the Fall as being a failure of memory—thinking humans have forgotten their primordial self. In a very real way they are right in believing the problem lies with forgetting, but this *forgetting* is not a failure of our sense of memory, rather, it is a failure in how we interpret memory. In other words, we have lost the clear, immediate experience of memory, and we have forgotten how to recover this experience. And worse, we don't even understand what we are missing, just a vague sense that we have forgotten something important. The most common misperception infecting both Eastern and Western philosophy is the thought that there is an inherent problem with the experience of phenomena itself. However, the actual problem is the cultural view imposed on reality. Memory is not by nature distorted; rather, the distortion is a consequence of our interpretation that we have impressed on memory. Certainly Parmenides recognized the problem was a misinterpretation of the phenomena, not a mis-perception[11]. For Parmenides, Truth (*Aletheia*) and Memory were synonymous[12].

The idea that the Fall is a fall into the temporal realm has often been pointed out, but the exact cause of the fall is shrouded in myth and silence. Several have remarked on the fall into time, but it was Swiss essayist and philosopher Jean-Jacques Rousseau (1712–1778) who was the first to directly recognize it as a fall into time: "Thus, Rousseau ... defined the central problem of human life as the discovery of strategies to correct, as much as it is humanly possible, the deformation of life brought about by the 'fall' into time."[13] Many others have commented on our involvement with time and noted that at the beginning there is timelessness, which subsequently descends into the temporal. The contrast is between the timeless Garden of Eden and the dust and noise of temporality. The advent of the fall divides the divine from the mundane, and once fallen we are caught in time, trapped by the way we think about things. Most often we never even suspect there is a way of thinking our way out of the human condition. As Wittgenstein put it, "How does the fly think its way out of the fly-bottle?"[14] Isn't our own entrapment because of the way we think? Doesn't the question become 'what's wrong with memory?' Jorge Luis Borges expresses it best, "Do you believe that the Fall is something other than not realizing that we are in paradise?"[15]

Religion and philosophy have two essentially different approaches to healing the schism that divides *man* from *nature*, and *self* from *other*. The problem, however it is to be resolved, is always a problem of time, because being born into time is an automatic sentence of 'death *by* time'. The religious solution offers an unlimited extension of time through a promised afterlife and in addition gives meaning and purpose to life by claiming that God cares, and it is the will of God. Philosophy, on the other hand, tries to either abolish or accommodate temporality, while striving to find purpose in the present. Judged on the basis of popularity, religion is far and away the winner. But religion offers up a paltry existence, unwilling or afraid to examine the extent of the delusion that we are condemned to wander in. The fog in which we wander is the fog of time and there is only one way out, and that is by solving the problem of time. Religion claims to have solved this problem by offering eternity, but if the problem is time, promising an infinite amount of time may not be the solution.

Generally speaking, the problem *is* one of time, but specifically the problem becomes a problem of understanding memory. Some mid-twentieth century philosophers arrived at a consensus that our understanding of memory amounted to "The Representative Theory of Memory."[16] This theory contends that all knowledge of the past "is based ultimately on memory"[17] which again points to the assumption that memory represents past experience. This would appear to be an obvious statement of fact, because memory is believed to represent a past time. Importantly, this illuminates the problem that the only empirical evidence for the existence of time is our experience of memory, assumed to be 'representative of past'.[18] To elaborate, the *future* is not empirical evidence of time because it is forever beyond experience, and the *present* without some evidence of past or future can not, on its own, be evidence of temporality. Thus it becomes clear that memory is the evidence on which time stands or falls. Undoubtedly this is what the ancient philosophers and thinkers sensed when they inquired into remembering and forgetting. Understand memory and you solve the problem of time.

For religion, the schism occurs because there is a separation of humans from God. In religion and also in all the ceremonies of all primitive societies there is a storied or mythical time when this divided state didn't exist. This mythical time, when there was a oneness with God, ended with what has come to be called the 'fall from grace' which according to the Bible was precipitated by taking the fruit of the Tree of Knowledge. The allegorical fruit suggests that knowledge is poison, thus producing the alienation from God. And this is the divide between religion and philosophy. Religion adopts a willful ignorance by choosing to believe without knowing, whereas philosophy wants to look for the nature of the poison by looking into the depths of our own ignorance.

To the philosopher this poisoned knowledge is seen as the root of the problem. The schism created can be healed only through reason, and by this, human beings can come to apprehend phenomena in a way that reveals reality. This should raise the question: What if the nagging sense of absurdity that seeps into our lives is the consequence of this very assumption about memory representing the past? If we can't assume memory to be a representation of the past, then time itself becomes a fiction, because the only empirical evidence for time is *presence*, which is no time at all. Thus memory is just another *sense* of the present, albeit, that mysterious *sixth sense*. What is it when memory is always now?

Chapter Three

Einstein and Piaget

In my beginning is my end.
—T. S. Eliot[1]

Discussions between Swiss psychologist Jean Piaget and German physicist Albert Einstein at a 1928 international conference on philosophy and psychology in Davos, Switzerland greatly influenced Piaget's ongoing child-development research project.[2] Piaget was seventeen years younger than Einstein, but at the age of thirty-one, had already established himself as an authority on childhood psychological development.[3] In their conversations, Einstein posed some incisive questions regarding the child's conceptual development of time and motion. The central question he asked of Piaget: "Is our intuitive grasp of time primitive or derived?"[4] Or to put this question another way: Is our grasp of time an innate human perception or is time something we learn as a child? After many years of research, Piaget, in 1946, published his emphatic and revealing answer in *The Child's Conception of Time.*[5] He and his research team were frustrated in their early investigations because, "... we quickly discovered, the time relationships constructed by young children are so largely based on what they hear from adults and not on their own experiences."[6] These preliminary findings should have been seen as a strong indication that the child's time conception was culturally derived from their adult teachers. Piaget's research went on to discover that time is a very complicated fabrication that is intimately tied to the autobiographical memory of the self, and in answering Einstein's question he convincingly demonstrated time is not primitive; rather, it is learned from our culture. Some who

19

came after Piaget quibbled about the details of his methodology,[7] and his discoveries have been further refined by others, but his conclusions are considered still valid today. The *Encyclopedia of Time* asserts " ... his [Piaget's] general conclusion stands: Time concepts are gradually constructed in the course of development."[8]

Psychologist William J. Friedman's survey, *About Time: Inventing the Fourth Dimension,*[9] published in 1990, notes Piaget's work and then goes on to compile the findings of psychological research on time since Piaget. Friedman asks the question: "Is time an automatic, fundamental property of memory?"[10] And in the next paragraph he answers: " ... we have found no evidence for its [time's] existence. Time memory seems more a makeshift affair, requiring a patchwork of processes and considerable effort—and usually ending in imprecision."[11] Since Friedman's publication, there is nothing new in the psychological literature contradicting the conclusion that time is a culturally instilled concept. However, research resulting from the coordinated efforts in the cognitive sciences over the past two decades has contributed to an ever more detailed story of the intimate connection between time, memory, and self-ego.

Piaget observed that our understanding of time is learned over many years in our early childhood. The child finally seems to 'get it' when she acquires the ability to tell stories about herself to herself, and to others.[12] At this point, the child's autobiographical memory has booted-up and is operational. The ego has been enabled, and the *being* that was prior to the conception of time has been captured by the temporal ego. Piaget's clinical observations provide us with a coarse description of the fall into time.

His research program in developmental psychology was an examination of how the self-ego is constructed, but to understand this he needed to discover the origin of time in the young psyche. In the end, he found himself going back to the beginning (i.e. earliest childhood) where time and self form into ego. We live in a cause and effect world

20

(the temporal world) so we naturally look to the beginning for an explanation—what has come before is thought to be the cause of our present circumstances. But what Piaget found, back at the beginning of our psychic development, is a point, before which time does not exist. This early condition of the very young child is referred to as *childhood amnesia*[13]—more about this in a moment.

Piaget examined memory as it develops over the time of childhood and then documented the conception of the self-ego as it unfolds in later childhood. Explicit in his studies is the fact that there is an entangled association between time, self, and memory. There comes out of this association of 'self and memory' an insight that has the force of an axiom: *We are what we remember ourselves to be.* And coupled with this is a corollary: *The self is constructed out of memory.* The idea that the totality of our memories is identical with what constitutes our psychic selves has come to be known as the *memory theory of identity.*

Piaget's research was confirming empirically what Enlightenment philosophers John Locke (1632–1704), E. B. Condillac (1715–1780), and Denis Diderot (1713–1784)[14] had previously stated. Historian Jerrold Seigel has condensed their thoughts into the statement: "The center that constitutes the self has a particular composition: it is memory."[15] Sixteen centuries before Piaget's psychological investigations, that great, early philosopher of time, Saint Augustine (354–430), was one of the first to declare the 'memory-mind-self' identity: "Great is the power of memory, a fearful thing, O my God, a deep and boundless manifoldness; and this thing is the mind, and this am I myself."[16] But it was British Enlightenment philosopher John Locke (1632–1704) who provided the detailed argument that shows the identity of memory, self and time. Philosophers Shaun Gallagher and Jonathan Shear explain: "Locke's solution was that consciousness maintains its identity over time only so far as memory extends to encompass past experience."[17] American philosopher and pioneering psychologist William James (1842–1910) writes that Locke's description of 'personal identity' "is just your chain of particular memories."[18] Psychologist Daniel M. Wegner notes that, "The essence of personal identity is memory. This was John Locke's (1690) view and has since formed the basis of several further developments of the *memory theory of identity*."[19]

A brief description of early child development would include the explanation that the years up till about age three or four are characterized by a lack of self identity,[20] and these earliest years define the period known as 'childhood amnesia.'[21] Interestingly, psychiatrist Ernest G. Schachtel "treats childhood amnesia as a symptom of lost innocence, as if it were our own fault—and our greatest misfortune—that we cannot hope to return to the Garden of Eden of our infancy."[22] The years after age three or four begin the attachment of memories to the self-concept[23] and thus begins the formation of an ego that extends into the past. This important development is concurrent with a language that is being acquired to handle these new concepts of the self and past. It is the case that " ... many modern memory researchers believe it [childhood amnesia] occurs because of a lack of sophisticated mental abilities, such as language, which are used to cue memory."[24]

Later chapters will detail how language plays a fundamental role in creating the self-ego through telling the story of the self-ego. The entire transformation from selflessness to self-ego is part of the same transformation from timelessness to time, and this change seems to hinge on the single supposition that *memory is past*. Piaget uncovered something at the root of the way we think—it is something we had forgotten, causing us to forget who we really are. The entire psychological foundation of time is nothing but the presupposition that *memory is past*. In the following text the italicized term *memory is past* will be used as an abbreviation for the fundamental presupposition that is responsible for our sense of time.

The exact influence Jean Piaget's work had on Albert Einstein's thought is unclear but Einstein, late in life, considered time to be an illusion.[25] Einstein's best friend in later years, Kurt Gödel, perhaps exerted a greater influence. Both Einstein and Gödel were at the Institute for Advanced Studies at Princeton, where Gödel demonstrated mathematically that if the Theory of Relativity was correct (and there has never been a scientific observation that has violated the theory) then time is a fiction.[26] Interestingly, Gödel, late in life, remarked that "*time* ... remains, even after Einstein, *the* philosophical question."[27]

Chapter Four

Entropy, Time's Arrow, and The Past Hypothesis

Humans die for this reason, that they cannot
join together the beginning and the end.
—Alcmaeon, 5th century BCE[1]

The experience of time is often said to have the quality of 'flow'[2] and this flow, in turn, has a sense of direction. Time, if it exists at all, must have some demonstrable physical property that imparts direction. If not, it would be difficult to claim there is any aspect of time that is other than imaginary. The various arguments put forward in support of the reality of time tend to be of two types—the psychological/linguistic and the physical/empirical.

Psychological/Linguistic Argument: Child psychologist Jean Piaget and the many later investigators have demonstrated that our psychological sense of time is learned, and this pervasive feeling of 'flow' persists in large part because our language is permeated with inescapable temporal implications.[3] Short story writer, essayist and poet Jorges Luis Borges prefaced an essay, which attempts to refute time, with the acknowledgment that "our language is so saturated and animated by time that it is quite possible there is not one statement in these pages which in some way does not demand or invoke the idea of time."[4] Many philosophers and thinkers approached the problem of understanding time through linguistic analysis. If our distinctive feeling of time's passage is deeply embedded in language, then linguistic analysis might reveal something about the nature of time. Philosopher John McTaggart in 1927 offered a detailed linguistic analysis of two common ways that we express time in language.[5] He noted time was commonly expressed in terms of *before* and *after*, or in terms of *past, present*, and *future*. The conclusion of his analysis was that these two methods of expressing time were logically incompatible and that time itself was unreal.[6] More than eighty years after McTaggart's publication, his arguments are still fervently debated by philosophers. Our psyche is so closely enmeshed in our use of language that it is impossible to extract a clear understanding of time from language alone.

Linguistic analysis of time seems locked in a logical loop. Because language presumes time at a very primitive level, words themselves limit a deeper inquiry into the reality of time. The temporal has already been assumed at a more fundamental level which lies below our consciousness. The assumed time, in the form of *memory-is-past*, is nearly pre-linguistic—it occurs back at the beginning of language. Even after all the effort that has been put into the linguistic analysis of the temporal sense, one is still left wondering about the true nature of time. Linguistic analysis certainly never discloses any evidence that time is real. For all practical purposes, Piaget has already solved the problem for us—there is no psychological or linguistic evidence for the reality of time—it is just a consequence of our culturally instilled way of thinking.

Physical/Empirical Argument: A far more promising source of evidence that would demonstrate the reality of time lies in the empirical arguments of physics. The physical/empirical search for time actually becomes a search for the 'arrow of time'.[7] The arrow of time is what gives direction to the sense of 'flow'—there is a sense of time 'passing' and the arrow of time indicates the direction of passing is from the past into future. In physics there must be some demonstrable direction to time or there is no time at all. If you cannot measure a specific direction of time it becomes, by definition, empirically impossible to even detect time. Unfortunately, for those who are seeking a measurable temporal arrow, all of the basic laws of physics are time-symmetrical.[8] As cosmologist Brian Greene said, " ... the laws of physics that have been articulated from Newton through Maxwell and Einstein, and up until today, show a complete symmetry between past and future."[9] The physical laws all run forward and backward with equal efficiency. The way the plus or minus sign for the time coordinate is assigned is strictly by convention, such that the arrow of time is purely arbitrary. Both quantum mechanics and relativity theory are essentially time-independent theories. However, there is a strong case to be made that time is observed in statistical mechanics, specifically in the Second Law of Thermodynamics (hereafter Second Law).[10] It is here that we find the concept of entropy, which offers the best case physics can make for an 'arrow of time'.[11] Probably the most prominent role the entropic arrow of time plays is in the current Big Bang model of the universe. Cosmologists have described their model as a universe that began 13.7 billion years ago as a consequence of the Big Bang singularity, which subsequently evolved physically through time up to the present-day universe.

A review of how the Second Law and entropy came into being will disclose the logic behind thinking that entropy clearly displays an arrow of time.[12] The common usage of entropy has come to mean the tendency of everything to, in time, collapse into disorder and chaos. We can never keep up with the increasing disarray; the dust accumulates in our lives and our bodies deteriorate over time. This common meaning is taken from the scientific theory of the Second Law, and embodies the consequences of the concept of entropy. The Second Law states, "the entropy of an isolated system strives toward a maximum."[13] Entropy, over time in a closed system, always increases.

In the nineteenth century, physicists understood that everything could be known about a gas if every gas molecule in a closed system could be measured for position and momentum at the same instant. Then the past of the system could be fully known and all future states of the system could be predicted. In fact, it was thought if the position and momentum of every particle in the universe could be known at a particular instant, then it was theoretically possible to reconstruct the entire past of the universe and also predict the entire future of the universe. Sean Carroll said it succinctly, "If we knew the precise state of every particle in the universe, we could deduce the future as well as the past."[14] This is the nature of non-statistical mechanics—it is time symmetrical and so there is no preferred direction of time. Cosmologist Victor J. Stenger said, "Indeed, the whole notion of beginning is meaningless in a time-symmetric universe."[15]

With proper empirical measurement of the present state, the entire past and the entire future can be known. However, the problem arises that it is simply not possible to simultaneously measure the individual position and momentum of even a small number of gas molecules in a closed system. An alternative is possible, which entails measuring the *average* momentum of the particles in the system. This is where, for the first time statistical theory enters into modern physics. Applying statistical theory required two assumptions—the assumption particles behave randomly and the assumption that each of the multitude of microstates-states of the system are all of equal probability. Averaging across a closed system of gas molecules worked well for establishing reliable relationships between temperature, pressure, and volume, and this gave physicists confidence that statistical methods are valid and have a wider application.

Notice, however, what is being lost. The process of statistical averaging does not preserve information about the position and momentum of each individual molecule—it is averaged out of the physicists' equations. This information (position and momentum) of the individual particles, taken collectively, contains all possible knowledge of the system, both past and future. In theory, if the microstates of all the particles at a particular instant are known, then classical physics can predict how the entire 'future' will play-out, and also postdict the entire 'history' of the system.[16] This is exactly the temporal symmetry that characterizes classical mechanics—time runs equally in either direction—the assignment of an arrow of time is purely arbitrary. However, the very exact particulate information is now lost through the averaging of statistical mechanics.

Effectively, statistical mechanics has averaged-out the detailed information about any thermodynamic system that is being studied. Statistical methodologies sometimes lead otherwise cautious scientists to misinterpret the data. And sometimes the data can be statistically manipulated to support the desired argument. It has almost become a maxim of current thought that people use statistics to lie. As Mark Twain (1835–1910) remarked "there are three kinds of lies: lies, damned lies, and statistics." Nevertheless, interpreters of this statistical theory, in the form of the Second Law, make the mistake of thinking that the theory can somehow predict 'a high entropy future,' and by extension, infer the 'low entropy past' that is necessary to Big Bang cosmology. Mathematician and cosmologist Hermann Bondi correctly captured the error in logic when he said, "It is somewhat offensive to our thought to suggest that if we know a system in detail [using classical mechanics] then we cannot tell which way time is going, but if we take a blurred view, a statistical view of it, that is to say throw away some information, then we can ..."[17]

Now, let's examine the consequence of the two assumptions inherent in statistical mechanics. First *random particle behavior* is assumed, and second *all microstates are of equal probability* is likewise assumed. Look closely and you will see why so many are fooled into thinking that entropy demonstrates an arrow of time that is directed toward the future. As stated in *The Encyclopedia of Time*, " ... irreversibility is an illusion produced by large numbers, but the real world, including our perception of it, is much more complicated."[18]

Mathematical probabilities that are grounded in observation seem like a good bet to rely on for predicting outcomes. And if the mathematical probability is extremely high, it would seem like a very reliable predictor of the future. Most thermodynamic systems involve an enormous number of particles, all of which are assumed to be moving randomly, and statistical analysis calculates that out of all the possible microstates-states (each microstate assumed to be of equal probability) the overwhelming majority of microstates will be at or near maximum entropy. So it would seem a safe prediction to say that any system which is at present in a 'low entropy state' will (and we are presently in a low entropy state, according to physics), at *any* later time, almost certainly be in a higher entropy state than the current state. Thus the Second Law of Thermodynamics is a statement that "the entropy of an isolated system strives toward a maximum."[19]

The Second Law is thought capable of predicting the future, so can it also, like classical mechanics, describe the state of a system at any time in the past? The answer is 'No', and here's why. The present day universe is in a state of low entropy so it is easy to see that if all the particles are moving randomly and all microstates are of equal probability, then at any *later* time it is statistically nearly certain entropy will increase. But, the same statistics when applied to the same randomness and same probabilities will likewise conclude that at any *earlier* time it is statistically nearly certain that entropy would have been higher. Thus, the Second Law predicts entropy will be higher in the future, but also predicts, with equal confidence, that entropy was higher in the past.[20] In no way can the Second Law discriminate the past from the future.

Columbia University physicist and mathematician Brian Greene said, "From any specified moment, the arrow of entropy increase points toward the future *and* toward the past. And that makes it decidedly awkward to propose entropy as the explanation of the *one-way* arrow of experiential time."[21] This is counter to the cosmological arrow of the Big Bang model, which insists that entropy was extremely low at the beginning and has increased ever since, and will continue to increase into the future until the final heat-death of the universe, at or near maximum entropy. The statistical implication that entropy was also higher in the past has forced cosmologists to add on another assumption that is contrary to the Second Law—the assumption that entropy was *lower* in the past. This assumption is called the *Past Hypothesis.*

Theoretical physicist Sean Carroll offers a concise explanation of the current state of the Big Bang Theory and how time and entropy fit into modern cosmology. He says, "The most mysterious thing about time is that it has a direction,"[22] and this direction [the arrow of time] is attributable to the many irreversible phenomena in the universe.[23] These irreversible processes are understood through the concept of entropy and it is also through entropy that the Big Bang model of the universe is understood.[24] Finally, though the Second Law always predicts a future increase in entropy, it can never explain the very low entropy required for the early universe, nor can it explain the present-day state of low entropy. The inexplicable low-entropy at the beginning of the universe necessitates an assumption called the *Past Hypothesis.*[25] Carroll accepts that there are serious problems with the Big Bang model, but makes it clear that he thinks the *Past Hypothesis* is absolutely necessary or our worldview will collapse.[26] However, physicist Victor Stenger offers a counter argument that not only is a timeless universe in agreement with the empirical evidence, but it provides a simpler explanation of reality.[27]

The majority of cultures devise a creation myth that tells the story of the beginning and for Western scientific culture that story, at present time, is the Big Bang model of the universe. It states that it all began 13.7 billion years ago out of a singularity—a point of indeterminacy. It was an event so improbable as to not even be believable, but most modern cosmologists do believe it. Physicist, mathematician and philosopher Roger Penrose has calculated just how unbelievable the Big Bang model is. Using the entropy concept of the Second Law to calculate the probability of the occurrence of the Big Bang, Penrose ranks it as **one chance** in 10 to the 10th power to the 123rd power.[28] This is a number so large it could never be written out without scientific notation. Penrose goes on to say it is a number that has more zeros, to the left of the decimal, than all the elementary particles in the known universe. It is indeed a strange paradox that this same statistical theory, by employing the entropy concept, predicts that the next consecutive moment in our universe will most certainly have a higher entropy than does this present moment, and which, at the same time, employs the identical statistical reasoning that makes the Big Bang model so improbable as to be unbelievable.

There are other hypothetical arrows of time, and in one way or another, these either appear because of the statistical fallacy of assuming randomness or because of the psychological fallacy of thinking that *memory is past.*[29] The statistical fallacy occurs as a consequence of trying to smooth-out, through averaging, the necessary individual measurements, but this averaging, without closer analysis, gives a superficial illusion of an arrow of time. The Second Law (because of the entropy concept) is undoubtedly the best candidate to demonstrate a physical arrow of time, but in the final analysis, there is nothing about the entropy concept that supports an arrow of time . In fact, it is safe to say there is no known physical/empirical arrow of time.

Statistical mechanics of the Second Law throws out all historical information and then claims it can predict the future. British philosopher Stephen Toulmin remarks, "Those who think of metaphysics as the most unconstrained or speculative of disciplines are misinformed; compared with cosmology, metaphysics is pedestrian and unimaginative."[30]

Chapter Five

Parmenides

To cure the work of Time it is necessary to *go back*
and find the *beginning of the world*.
—Mircea Eliade[1]

Fragments from a single poem are the only known record of Parmenides' thought. Although fragmented, "the overall outline of the poem is clear" and contains "the longest, most continuous, and most cohesive text from the period before the Sophists."[2] Written in the early 5th century BCE, the poem, *On Nature,* stands as the first systematic philosophical writing in the Western tradition and exerted a great influence on Plato[3] and all who followed. The poem is presented in three parts. The first part is a prologue which serves as an introduction to the two main divisions of the poem. These two divisions are presented as alternative paths, the first path being the Way of Truth[4] and the second the Way of Belief.[5] The Way of Truth describes the timeless unity of Being, while the Way of Belief describes the delusion of mortals.[6] The contrast between the two paths has been described as the difference in the view from the gods' timeless perspective versus the view from the mortal's temporal perspective. In the words of British classical scholar David J. Furley, "the Way of Truth is the way an immortal looks at the world ... the Way of Seeming [the Way of Belief] is the way mortals see the world in time."[7]

The prologue to *On Nature* is a narration of Parmenides' own philosophical journey,[8] told in the form of an allegory. The poem opens with the young Parmenides riding in a chariot driven by two goddesses—the Daughters of the Sun. Parmenides is leaving the 'House of Night' traveling to the light[9]. The chariot arrives at the bolted gates of the paths of Night and Day, guarded by the goddess Justice. The charioteers' calm reasoning with gentle words convinces Justice to open wide the gates.[10] The chariot proceeds on the broad way so the young Parmenides can meet with *Aletheia,*[11] the Goddess of Memory.[12]. Aletheia receives him kindly and takes his hand in hers and she begins to speak the words that narrate the remainder of the poem. The Goddess of Memory tells Parmenides he has traveled a path that is far from the beaten path of man. For his quest Aletheia promises to tell him the heart of the Way of Truth and also explain the illusion which is the Way of Belief that mortals have fallen into. This concludes what has come to be called the prologue of the poem.

The text regarding the two Ways—the alternative paths described by Aletheia—is mostly a description of the two different circumstances that prevail on the alternative paths. Martin Heidegger said "The goddess of Truth who guides Parmenides, puts two pathways before him, one of uncovering, one of hiding ..."[13] Parmenides has an "obligation to choose between the world of *being* and the world of *opinion.*"[14] Most commentators/interpreters generally agree on what is being described in the words of the poem, but there is commonly disagreement about the overarching purpose and meaning of the poem. In particular, with the prologue there is general agreement about what is being described, but divergence about the *meaning* of what is being depicted. Broadly speaking, two sorts of interpretations are generally imposed on the poem. The first argument suggests that what Parmenides is saying must be more or less continuous and compatible with the Archaic literature of Hesiod and Homer who preceded Parmenides, and/or compatible with the Pythagoreanism that was contemporaneous with Parmenides. Alternatively, others tend to think that Parmenides is speaking in a way that no previous Greek had spoken—and what he is saying is profoundly different from what came before.

There is no reason to doubt that Parmenides is a unique thinker.

Likewise, there is little to recommend the interpretation that he is a throwback to the earlier thinking of the Archaic period of Greek culture and there is little in the poem to connect him with Pythagoreanism.[15] The fact that Parmenides was so original argues strongly for the idea that he was saying something completely new.[16]

A short review of Parmenides' accomplishments will illustrate just how creative and original he was. Based on the historical record, he was the first to attempt the construction of a systematic philosophy. He was the first to report four important scientific discoveries, namely, "that the earth is a sphere; that the two tropics and the Arctic and the Antarctic circles divide the earth into five zones; that the moon gets its light from the sun; and that the morning star and the evening star are the same planet."[17] Additionally, his is the first recorded allegory, he is also the founder of epistemology,[18] and he greatly impressed the later Greek philosophers. In all the *Dialogues of Plato*, Socrates never defers to anyone—except Parmenides.[19] And the greatest influence on Plato's philosophy was Parmenides.[20] Western thought was changed by this poem written 2,500 years ago.[21] Parmenides said something uniquely original and most who have examined the existing 150 lines of his poem believe his message has also been preserved. What was it that Parmenides actually said?

Most of the debate about the meaning of the poem has centered on Parmenides' descriptions of the Ways of Truth and Belief, but the key to understanding his actual intent is contained in the Prologue. The obvious interpretation seems to have been overlooked. Parmenides was indicating in every possible way that the problem of humans was with memory. The overarching meaning of the poem is that the Way of Truth *is* the Way of recalling our timeless memory, while the Way of Belief is the fall into time. Seen in this light, the Prologue is pointing directly at restoring pure primordial memory as the key to Truth.

In late Archaic Greece, Aletheia was a goddess whose name literally meant 'not-forgetting'. Pure uncontaminated memory was once synonymous with Truth, but by the beginning of the Classical Period the term Aletheia itself had fallen into forgetfulness and its meaning had been lost. French historian and anthropologist Jean-Pierre Vernant is adamant that the Archaic Greeks perceived memory quite differently from those who followed. He said, "Memory appears to predate any consciousness of the past and any interest in the past as such."[22] Heidegger, speaking of the earliest Greeks, likewise tells us that "*Memory* initially did not at all mean the power to recall."[23] The term *Aletheia*, by the Classical period, had devolved into just a word that meant truth, but the Truth itself had been lost. Socrates knew that he didn't fully grasp what Parmenides was saying. Socrates said of Parmenides, "I am afraid we might not understand his words and still less follow the thought they express."[24] The goddess's name was still spoken, but the intended meaning was forgotten. And by the time of Plato, the forgetting was nearly complete—Truth was forgotten. Plato had retained some vestiges of Parmenides' insight—" ... remembering or recollection ... is the basis of Plato's theory of knowledge and wisdom."[25] But the understanding of the nature of memory had gone into hiding. It has been said, "The beginning cannot be preserved as beginning; it can only be remembered or forgotten."[26]

Since antiquity there has been a long-running philosophical discussion about what Parmenides meant by Aletheia (Truth).[27] It is well understood that the poem is describing two distinct, alternate paths: The Way of Truth (Aletheia) describes a timeless, changeless, unity of all-in-One, and the Way of Belief (Doxa) refers to the mundane world, where a confusion of presumptions has been taken on as beliefs. To know the Way of Truth requires understanding Parmenides' Maxim, which states 'what *is*, is, and cannot not be.'[28] Parmenides stated and restated this maxim several times in the verses of the Way of Truth. A passage from the Bhagavad Gita, written in India during the same era, parallels what Parmenides is expressing: "Of the non-existent there is no coming to be; of the existent there is no ceasing to be."[29] Parmenides is saying 'Being is', and non-being is impossible, and this makes reality a timeless, changeless, all-in-one. Welsh philosopher G. E. L. Owen said, "Surely, we are inclined to say, if non-existence is ruled out then there *is no such thing* as change? And then, equally surely, what there is must stay the same?"[30]

Philosopher Karl Popper sensed that there might be a reluctance on the part of the historians of science or philosophy to attribute to Parmenides such a radical idea that 'change is an illusion.' Popper said they might "perhaps be less reluctant when they see that great scientists, such as Boltzmann, Minkowski, Weyl, Schrödinger, Gödel, and above all Einstein, have seen things in a similar way to Parmenides, and have expressed themselves in strangely similar terms."[31] The unchanging unity that Parmenides was describing prompted physicist Erwin Schrödinger to say, "His 'truth' was the purest monism ever conceived."[32] Austrian-born philosopher of science Paul Feyerabend said, "He also anticipated a popular interpretation of the theory of relativity ... "[33] Feyerabend is referring to the idea that Parmenides' thought prefigured the "Block Universe"[34] interpretation of Einstein's spacetime continuum, in which the appearance of time is an illusion. There is so much that Parmenides revealed in the extant fragments of *On Nature. S*till, something important regarding Parmenides' description of truth remains deeply hidden.[35]

Martin Heidegger, who has dug as deeply into Parmenides as any modern philosopher, said "The field of the essence of aletheia is covered over with debris."[36] But even in Parmenides' time Aletheia was being covered over by cultural changes that started with Simonides[37]. The Prologue of the poem is a mnemonic device, bringing to mind that Memory and Truth are identical. The Prologue is revealing the switch that operates between truth and belief, and it all comes down to a single belief—a belief about memory. To recognize this single belief is to remember what has been forgotten. We forget reality when we believe that *memory-is-past*, and because we have *forgotten* direct perception of memory, our worldview is profoundly altered. The Goddess of Memory is bringing to mind *not-forgetting*—the literal meaning of *aletheia*.

The Way of Truth focuses on another mental error that involves 'thinking into existence' the *being* of the non-existent. Parmenides' Maxim, repeated throughout the Way of Truth, is a caution against believing in what is not-present, which results in memory being misinterpreted as representing the past. But *Truth* is the unmediated present memory—we "remember" in the present moment. For Parmenides, 'not forgetting' (Aletheia), is unadulterated memory. In the end we arrive at the beginning—as Parmenides said: "It matters not at all where I begin; I will return again to the same place."[38] Heidegger said, "... *aletheia* is the beginning itself ... The thinker thinks the beginning insofar as he thinks *aletheia*. Such recollection is thinking's single thought."[39] Parmenides' *On Nature* is one of the rarest of writings—it not only reveals that the human 'fall' is into time, and that time *is* memory, but it also reveals that memory is not time.

Parmenides is a true enlightenment philosopher who is preaching the same enlightenment as the Buddha and Lao Tzu. Parmenides, as much as anyone, was a man of the Axial Age. He knew we had forgotten ourselves and that it was a defect of memory that caused the fall of those who follow the Way of Belief. Humans believe in an untruth about memory, and only the goddess Aletheia can bring back the truth of memory. The Way of Truth is the path of primordial memory, whereas the Way of Belief is the fall into time.

Chapter Six

Self and Meme

The self is formed out of a tightly wound, nearly impenetrable nexus of time-self-language-memory. What is obvious but often unacknowledged is that the most important component of the self is memory. Philosopher Edward S. Casey explains, "It is an inescapable fact about human existence that we are made of our memories: *we are what we remember ourselves to be.* We cannot dissociate the remembering of our personal past from our present self-identity. Indeed, such remembering brings about this identity."[2] This has come to be called the Memory Theory of Identity. If we are our memory, it is not surprising that the brain is thought to harbor the self.[3] After all, "Memory is not the domain of particular systems in the brain, but of the brain as a whole. Memory is what the brain does,"[4] and there seems to be almost an identity between brain, self, and consciousness.

Only humans are known to have a consciousness characterized by a clearly defined sense of self-awareness.[5] Other organisms experience consciousness of the environment but not self-consciousness. "There is reason to think that animals or babies can have phenomenally conscious states without employing any concept of the self."[6] The sense of personal identity and self-reflection/self-consciousness is clearly tied to language. Learning language is what makes possible the internal voice that we associate with the self. It is language that brings the child out of the state of childhood amnesia.

Modern attempts to specify the exact locus and nature of the self within the brain occurred as early as René Descartes, who believed that the self was centered in the pineal gland,[7] but since the time of Descartes, various mental functions have been found to be associated with specific areas of the brain. Recent research in the cognitive sciences[8] of the past few decades has developed a more concise picture of the *self* that lives in our brain. There is now an understanding of the general architecture of the brain, and much has been learned at the cellular/biochemical level, along with some rough idea of the wiring scheme. We have fairly good knowledge of how information is transferred and stored—'thinking' at the 'machine level' of the brain approximates a cybernetic function. The human brain has been referred to as three pounds of wet-ware analogous to the hardware of a very sophisticated computer.

The remembered self is obviously composed of the stored memories of experiences, but more importantly, the sense of the self and self-consciousness is derived from the program that has been written onto/into memory. The programming of the self, which has become a part of the human brain's operating system, is a consequence of the culturally installed software program. Our operating system might be predominantly hard-wired, but the portion that creates the self is imprinted on our memory by the culture in which we are raised. Philosopher Daniel Dennett has remarked that there is sufficient understanding of how the hardware works—what we need to know is what is the nature of the software that is directing our thinking.[9] Recent empirical observations of the brain combined with insights from the science of memetics[10] (more about memetics in a moment) offer a convincing case for how culture encodes the *self-program* into our memory and how this program then interprets phenomenal experience so that it appears to be the self-experience in time and space.

There are essentially two categories of long-term memory: semantic memory and autobiographical (or episodic) memory. The semantic memory is all the facts that are known but are not associated with time or place.[11] These are memories such as knowing that $2 + 2 = 4$. We know that we know it, but there is no recollection of having learned it. The autobiographical memory comprises all the memories associated with a time or place and are memories most closely associated with the self. As John Kotre points out, "Besides recall, something else is needed for autobiographical memory—a sense of self. Babies are born without one. They possess neither *I* nor *me*, neither the self-as-subject nor the self-as-object."[12] It is the existence of the autobiographical memory that provides a convincing argument for the existence of a self, because we *are* that which we remember ourselves to be. To gain further insight into this self-consciousness and how it is constituted out of memory, time, and language, it is important to enquire into what memetics has to say about the problem of the self.

The theory behind memetics was first suggested by Richard Dawkins in his 1976 book *The Selfish Gene*. He proposed that there were *units of information*, which he called memes (rhymes with *seems*), that were transferred, commonly through language, from the brain/memory of one person to another. These memes behaved very much like genes—if they replicated they 'lived' and if they didn't they died out. By the early 1990s, the term *meme* had entered the popular culture and had taken on a meaning somewhat similar to what Dawkins intended. Popular culture removed some of the sterility of the term *meme* being defined as a "unit of information": it had come to mean any fad, habit, story, tune, saying, cliche, or "any other kind of information that is copied from person to person."[13] Memes were moving through culture passing from one person's memory to another.

The *meme* concept was adopted so easily because it seems intuitively correct that popular thoughts persist, and less popular thoughts die out. What Dawkins had hit on was also a very powerful way of analyzing cultural evolution and cultural drift.[14] Many scientists recognized this as an idea that could be exploited—the science behind this idea is called *memetics,* formed around the concept that human cultural evolution has units of information (memes) that behave very much like the units of information in genetics (genes). And it didn't go unnoticed by those scientists thinking about memetics that the meme often behaved very much like a cyber-virus, jumping from memory to memory.[15] Rather than infecting one computer after another, it was infecting one mind after another. Dawkins, speaking of the complex of memes that form our ideological belief systems, referred to them as "viruses of the mind"[16]. Note that "The vast majority of memes are not [harmful] viruses but are the very foundation of our lives ..."[17] However, *Time Sutra* is focused on the most fundamental meme, the meme that is the ground of our worldview, and a meme in position to do the greatest harm.

Many of those in the cognitive sciences realized that the theory of memetics offered a way of informing ourselves about both the nature of culture and the nature of the self. From this perspective on memetics, it quickly became apparent that the self was, in large part, made up of a complex of memes.[18] This meme complex forming the self is a composition of the total cultural/social and personal encoding. Within the complex of memes that constitutes the self, there is undoubtedly a firm belief that memory represents the past history of the self. Although this belief is not encoded as a single short statement specifying *memory-is-past,* this particular meme itself *does* reside in some form in the brain. And so the statement '*memory-is-past*' is used as a shorthand notation to designate this particular meme, which is in many ways encoded into memory.

Given our current understanding of the brain from knowledge of memetics and, more broadly, knowledge derived from the cognitive sciences, a model has emerged that describes how the *self* is fabricated out of a complex of beliefs. What follows is a rough sketch of how the brain is deceived into thinking that the self-ego exists in time and space.

Careful review of what had been taken to be empirical observations regarding time show that time is actually derived from a belief about memory. However, humans from all cultures, to one extent or another, have taken on the belief that time exists. The belief in time is undoubtedly taught by the culture in which one is raised. In Western culture it has been shown that the child learns the Western concept of time over several years. Not until the child can tell the stories of the *self* to himself and others does he fully grasp the Western concept of time. At this time the child has a sufficiently developed ego, and from this he goes on to become the fully enculturated adult member of society. The brain has acquired a self.

Whatever time is, we know that it is a complicated process to inscribe the program into memory and that time is not fully understood (believed) until there is sufficient development of the self. It is obviously not a coincidence that the self is made out of memory and that time is also made out of memory. Keep in mind, our only empirical evidence for time is the existence of *past,*[19] and the only evidence for past is belief in the meme, *memory-is-past*. By another name, this belief has been called the Representative Theory of Memory[20] and provides our direct reason for thinking that both self and time are real (the presupposition of past has also recently been identified in theoretical physics and cosmology, and appropriately termed the 'Past Hypothesis'[21]). Believing the meme, *memory-is-past*, has the identical consequence of believing the 'Representative Theory of Memory'. The belief creates the illusion of time, and unknowingly distorts the timeless-nature of memory into an unnatural temporality. This is a distortion of our nature and it brings on an illusion from which there is no easy escape.

The brain is made up of 80-100 billion neurons (brain cells), with each neuron having thousands of thread-like dendrites leading from every cell. The dendrites act as conduits along which potential impulses can be sent to other neurons, forming an astoundingly complex architecture. The brain has been said to be the most complex structure in the known universe, and many believe that understanding this structure is the greatest challenge in science.[22]

The forebrain, of very recent evolutionary development, occupies most of the brain's volume. At birth, the forebrain is a vast memory that awaits programming. But the brain at birth is already processing information from all the senses—this massive parallel processing is an architecture that modern computer designers would like to duplicate.[23] Certainly the information that is being processed by the brain, in some way, represents our consciousness, but we are never conscious of all the different senses at once—consciousness can only attend to one thing at a time. Much has been said about multi-tasking, but cognitive scientists say there is no such thing—there is only the ability to rapidly shift attention from one concern to another.[24] It is impossible to focus on two things at once. Consciousness (our attention) is always attending to one sense at a time. For instance, our attention may be focused on *watching* and a loud noise will divert our attention to *listening*, but we can't consciously do both at once. The brain is always actively, and unconsciously, processing the various data streams incoming from all the senses,[25] but our consciousness is focused on just one sense at a time.

Daniel Dennett has offered a compelling description of how this complex parallel processing takes place below the level of consciousness. Each sensory input is constantly and unconsciously being monitored while simultaneously there is a consciousness that can be aware of, at any time, a single stream of sensory input. The consciousness behaves very much like a linear processor (a von Neumann machine[26]) which has the ability to quickly switch from one data stream to another, but only one at a time. Thus consciousness can focus on what is being seen, then what is being heard, then what is being tasted, then what is being thought, etc., but always one sense at a time. All of this is descriptive of consciousness, but what of self-consciousness? Here is where language becomes important—we all have in our heads a self-narrative that acts as a verbal stream of consciousness.[27] This imposes a narrative overlay on conscious awareness that personalizes it and makes it 'my' self-awareness—it is the voice of the owner of the perceptions.[28]

This brings us to the six senses—five of the body and one of mind. Of course there are more than five bodily senses (touch is made up of perception of heat, cold, pressure, pain—all with different receptors), but it is convenient to stay with five traditional bodily senses of sight, sound, taste, touch, and smell. The single mental sense is most often called *mind*, but it should probably be called *memory*. In the text that follows, the perception of memory will often be referred to as the sixth sense.[29] The newborn is processing all incoming five senses of the body, but the sixth sense of memory is mostly a recorder that is turned on. The memory is intensely listening and recording the incoming programming, and the incoming programming is not only the experience of the five bodily senses, but memory is also immersed in the cultural programming that will form the socialized human being who is learning to believe in time.

The enculturation that leads to the belief in time comes from this indoctrination of the self. From birth, all those who surround and interact with the newborn treat her as a self. As she learns the language of her culture, she learns to speak as a self that is embedded in personal and cultural history. The culture doesn't know that it is teaching *time*—the teachers are as ignorant of the nature of time as the newborn. However, even the most primitive cultures teach *before* and *after*, and *past* and *future*.[30] Time is taught but there is almost no understanding that it is being taught. What is being taught is quite simple: the experience of memory is 'not-present', it is 'before-now'. The child grows into the adult unsuspecting that her entire experience is seen through a cultural lens about which she knows almost nothing.

The frame that makes the worldview is now in place. Whenever 'long-term' memory is present, it is seen as past. The meme '*memory-is-*

past' is placed in the position of a filter through which all 'long-term' memories must pass and essentially this entails all of permanent memory. The temporal terms *'long-term'* and *'permanent'* are being used to describe memory since that is how we have come to recognize it. We recognize it in this way because it has already passed through the filter which codes the experience of memory to mean 'before now'. We have learned to see our memory as history and we cannot experience it any other way. When we experience this type of memory we are having an experience of the *self-ego*. This is the origin of the self-ego: we are what we remember ourselves to be. There is nothing more to self-ego than this distortion of memory.

We have branded a segment of the memory as temporal, but there are still many operations of memory that are not labeled temporal and seem very much in the present—this is the experience of short-term memory and working memory. Although all thoughts are from language, and language is derived from long-term memory, many of our thoughts have a sense of immediate presence. When memory is not entangled in thinking that the present experience of memory represents the past, then there is an immediacy of being. This immediacy feels like present self, but this is a self with a freer sense of presence, because it is not anchored in time and history. This type of experience includes those ecstatic moments when the self has lost itself in timeless and selfless being. These experiences are direct insights into unmediated reality.[31]

Often the experience of insight is nothing but the self losing the self and returning to being. The experience of the self is nothing other than the experience of memory that has been screened through the *memory-is-past* filter. The great problem that arises from this kind of thinking is that it creates a self out of time that is burdened by greed, anger and ignorance[32], and it is mostly these negative qualities that stimulate the madness so common in human actions. The self that is made out of time is the author of all ideologies, whether religious, nationalistic, political, economic, etc.

With the temporal filter in place, the experience of memory becomes the experience of self-ego. The culture has spent several years getting this filter in place and closely fitted—such a close fit that we are unaware of the filter's presence, so there is no inclination to look at the unfiltered reality. The self-ego is completely engulfed in time, believing fully in the illusion that being is an ego living out a life of the past, present, and future.

Part II

Anamnesis

Chapter Seven

Archaeology of the Self: Language, Duality, Belief

... the act of anamnesis—an escape from time,
a revelation of immutable and eternal being ...
—Marcel Detienne[1]

Time Sutra is a meditation on time which leads inevitably to the contemplation of memory. This meditation on memory is intent on digging up certain artifacts of culture—those instilled beliefs that lie unnoticed below consciousness. Once *time* and the *self-ego* are shown to be nothing other than 'mental constructions', then the meditation on memory deepens. For all of the other five senses, we never assume that they are anything other than present phenomena. But if you want to experience the sixth sense (memory)[2] as *present* phenomena, then the belief that it represents something 'not present' must be exhumed and brought into the light. This meditation on memory is the practice of an 'archaeology of the self'—exhuming ever deeper layers that go all the way down to the beginning of time. Archaeology is the study of the remains of a culture. In this case, the remains are the memes that our own culture installs in memory. Anamnesis[3] (calling to mind knowledge that has been lost) is the method by which we delve into an archeology of the self—remembering what we have forgotten.

The self is assembled out of a matrix of memories, held together by thought and language. At the core of the self is the meme proclaiming *memory-is-past*. And it is around this core belief that language provides a narrative which forms the conscious experience of the self-ego.[4] To further investigate this structure of the self requires an understanding of

51

how language works. This in turn leads to the nature of duality and the consequences of belief.

Swiss linguist Ferdinand de Saussure (1857–1913) noticed that a word has meaning based on its difference from other words. Philosopher Hugh J. Silverman said, "Indeed, de Saussure defines a sign [word] as determined by its difference from all other signs in the sign system."[5] More generally, any symbolic system such as language is based on the differences between symbols, and the entire system is a closed, self-referential unity.[6] A good example of this self-referential 'system of differences' is the contents of an unabridged dictionary of any language. Every word within the dictionary is defined by other words within the same unabridged dictionary—all words refer ultimately to other words within the dictionary. Even though many of the words are symbols for experiences of the world (sights, sounds, etc.), or symbols for things *in* the world (trees, cars, etc.), words are never the actual experiences of the world, nor are words the actual things in the world. The words are only symbols. Any self-referential system of symbols, such as occurs in all languages (and in mathematics[7]) has come to be referred to as a *hermeneutic circle*[8]. It is a point of logic that the hermeneutic circle can say nothing directly about that which is outside the hermeneutic circle. Many see this hermeneutic reasoning as one more skeptical argument denying the possibility of true knowledge.[9]

Language is mostly limited to descriptions of the world—descriptions in words that are limited by their definition. The world is described by words whose meanings are defined in terms of other words. Words gain their meaning from their differences from other words. To say anything that isn't purely descriptive requires language to become very circumspect. Language must use allegory, metaphor, simile, and analogy to allude to a reality that is beyond language.

Language is how we think about the world, so it is critical that we understand how language is entangled in this core belief (*memory-is-past*) that makes our world a world of time. De Saussure's observation

52

that a word derives meaning through its difference from other words is an observation closely related to the way that dualities also function through difference—the two opposing terms of a duality give meaning to each other. American philosopher Richard Watson said dualism is "the view that reality consists of two disparate parts."[10] A duality, in the context it is being used here, is made of two opposing terms, often antonyms, expressing the antithesis of meanings. For instance, Love-Hate and Hot-Cold are both dualities of opposition. The first instance expresses opposite emotions about someone or something, and in the second instance the duality is describing opposite extremes in the range of temperature. We sometimes think in terms of extremes, but language supplies many intermediate terms that provide a more subtle shading to our meaning. In the case of Hot-Cold, the two extremes are in-filled by descriptive words that fall along the spectrum of difference between the two antithetic end-members. We refine the continuum into fiery, heated, warm, tepid, cool, chilly, freezing, along with a multitude of other modifiers describing differences of temperature.

All dualities ultimately emerge out of language, either in the simple form of words of opposition, or the more complex form of conceptual thinking that is made possible by language. These two types of dualities could be termed *linguistic* and *conceptual*. *Linguistic dualities* are a direct product of the 'differences' that De Saussure recognized as the primary function of language, whereas the *conceptual dualities* are derived from the thoughts that language makes possible. However, there is one conceptual duality in particular that has engaged the interest of philosophers, who often refer to this concept as *substance duality*. Substance duality is so named because it appears that reality is composed of two different and seemingly incommensurable substances—*mind* and *matter*. Richard Watson explains that the influential American philosopher John Dewey "... finds all the problems of modern philosophy derive from dualistic oppositions, particularly between spirit and nature."[11] Although substance duality has been the focus of many thinkers over a long period of time, it too, is readily explained as a consequence of the schism of the senses, resulting in the closely related dualities of mind-body, self-other, spirit-nature, subject-object, interior-exterior, and mind-matter. This seemingly incommensurable difference in the nature of the two opposing aspects that form the *substance duality* will be fully explained in the Chapter 9 discussion on The Explanatory Gap.

Much criticism has been leveled against dualistic thinking. However, as philosopher Richard Watson said, "despite the extremely difficult problems posed by ontological dualism, and despite the cogency of the many arguments against dualistic thinking, Western philosophy continues to be predominately dualistic ..."[12] Philosopher Daniel Dennett makes it clear that he feels this type of thinking is most certainly wrong when he says, "For many people, this idea (dualism) is *still* the only vision of consciousness that makes any sense to them, but there is now widespread agreement among scientists and philosophers that dualism is—must be—simply false ..."[13]

Even though all of our thought takes place in the form of dualistic thinking, the concept of duality, when first encountered, seems foreign. Understanding dualistic thinking requires that we know that any given concept depends for its meaning on an (often unspoken) antithetical concept of opposite meaning. Richard Watson said, "Dualism is related to binary thinking, i.e., to systems of thought that are two-valued ..."[14] Some common dualities that philosophers wrestle with are good and evil, truth and error, belief and doubt, etc.[15] The problem is that one side of a duality loses all meaning without its opposite, thus *good* has no meaning absent *evil*.

The black and white photograph provides an apt analogy of a dualistic opposition. Each light-sensitive granule that makes up the emulsion on the photographic film will either react or not react when exposed to light, so the exposed and developed photograph is composed of a fine matrix of black and white dots. Taken together, these dots form the photographic image. There can be no image without the white dots, just as there can be no image without the black dots—the two opposites are each equally necessary. Of course it appears that there are all gradations of grey in between, but they are contingent on the existence of the opposition that forms the black/white duality. The visual experience of the photograph is ultimately derived from the duality. All of this likewise takes place in the natural function of language—it functions because of innumerable dualities forming a matrix of relations/oppositions. However, there is one type of duality that is of particular importance to this investigation if we are to ever unveil the memory-time-language-self nexus, and this particular duality is formed in the relationship between belief and doubt.

Beliefs exert a strange power over how we think. Once a belief is accepted, it takes on an aura of truth. All the ideologies—religious, economic, political, etc., are powered by beliefs—by those ideologues who believe with complete conviction that they have found the truth. Richard Feynman, physicist and philosopher, observes that the worst of times in human history are "times in which there were people who believed with absolute faith and absolute dogmatism in something."[16] Once a belief becomes embedded in the self-ego, it is no longer visible to the mind. Beliefs that gain a hold over *self* are, in a way, a type of forgetting. The only possible recovery is by restoring these beliefs to conscious memory—through anamnesis. Philosopher Eric Voegelin said that you must "go as far back as ... remembrance of things past would allow in order to reach the strata of reality-consciousness that were the least overlaid by later accretions."[17]

Belief begins to go into hiding at the very moment belief begins. Once the gloss of a false truth covers over belief, the fact that it is a belief is lost to memory. Here begins the strange relationship between belief and doubt. Doubt appears along with belief, because doubt is not possible without belief.[18] Once belief appears doubt becomes possible, and if truth is to be sought then doubt becomes necessary.

To the believer, the duality appears as Truth/Doubt rather than Belief/Doubt, so there is never a serious questioning skepticism that can arise. As philosopher Ludwig Wittgenstein said, "The difficulty is to realize the groundlessness of our believing,"[19] and philosopher Karl Popper elaborates, "... we can discover the fact that we had a prejudice only after having got rid of it."[20]

We take on beliefs because they are so compelling when they come wrapped in an ideology. The ideologies promise wealth or power or eternity or happiness, so there is an enormous desire to believe. The most rigid kind of belief is *true-belief* because it will not admit doubt, thus what has become hidden—becomes lost. French philosopher Michel Montaigne (1533–1592) said it best: "There is a plague on Man: his opinion that he knows something."[21] Buddhist philosopher David Loy explains the incredible difficulty in allowing doubt to penetrate the bubble of true-belief. "How uncomfortable it is to realize that one's opinion of something is wrong and needs to be changed. How much more anxious does one become when one starts letting go of all one's opinions about the world and, most of all, one's opinions about oneself—to let go of the self-image whereby one's self is fixated."[22]

American computer programmer and author Richard Brodie, writing in *Virus of the Mind,* describes the problem from the perspective of memetics. Substituting the word *meme* for *belief,* he said, "Labeling a meme *True* lodges it in your programming and eliminates your conscious ability to choose your own memes [beliefs]. Once some authority convinces you something is True or Right or is something you Should do, you are effectively programmed."[23] Francis Bacon (1561–1626) recognized the belief/doubt duality: "if a man will begin with certainties; he shall end in doubts; but if he will be content to begin in doubts he shall end in certainties."[24] The great practitioner of skepticism, René Descartes, said, "If you would be a real seeker after truth, it is necessary that at least once in your life you doubt, as far as possible, all things."[25]

Realizing these deeply buried beliefs is a form of remembering, and systematically uncovering these beliefs is the practice of anamnesis. In a way, the hidden beliefs are never forgotten because they directly influence our present way of thinking. This way of thinking can never end unless we recover those earliest memories, back before we fell into time. Even our earliest memories were 'present memories' until we began to believe they were 'before now'—this is also at the time when we began to forget. This is how forgetting and remembering happens, and this is how we must, through a meditation of 'calling to mind', begin to recover that which has not been forgotten, just hidden from consciousness. Excavate all the memories that have been hidden by belief through the meditative process of anamnesis and you will expose the mechanism by which memories become hidden.

You cannot experience the nature of the self-ego without acquiring an inventory of the deep-seated beliefs that guide, if not control, your thoughts. Discovery of the founding beliefs of the ego is a particularly difficult task—how can you find something you don't even know exists? The truth is, all humans harbor a secret cache of unexamined beliefs. Fortunately, there are only a very few foundational beliefs that set us out on this disastrous path we have taken. The practice of anamnesis is the recovery of these few most basic memories from that moment when we fell into time. The promise of this practice is stated by the ninth century Japanese Zen Master Dogen: "Transcend discrimination of opposites, discover total reality, and achieve detachment. This is complete freedom."[26]

Chapter Eight

Sixth Sense: Memory and Mind

Mind is memory, at whatever level, by whatever name you call it ...
—Krishnamurti [1]

The term 'sixth sense' as it has been used in both East and West is plagued with confusion. Hannah Arendt reported that the first modern use of the term in the West was in 1730[2] and was used to denote 'common sense', which was in turn a phrase used since the time of Thomas Aquinas.[3] Aquinas intended his *sensus communis* to indicate a mental sense that unified and integrated the five bodily senses.

In the late nineteenth century in an atmosphere of seances, mediums and ouija boards, Sir Richard Burton coined the term extrasensory perception (ESP) as a catch-all phrase for any sixth sense of the mind that might exist. By the 1930s, psychologist J. B. Rhine had set up the Rhine Parapsychology Laboratory at Duke University, the first academic research facility to pursue scientific investigation of this mental phenomena. By the late twentieth century, discouraged by the lack of any consistently replicable experimental research, paranormal investigations had fallen out of favor on most American university campuses. The use of the term sixth sense has, in the West, continued to conjure up some kind of mysterious intuition or inexplicable ability. This impression was furthered by some of the esoteric Eastern writings that made their way to the West and fueled this sense of mystery.

However, the context in which the sixth sense is being used in *Time Sutra* is straightforward—the sixth sense is the faculty of *memory*. Declaring the sixth sense to be *memory* grounds it in a more empirical phenomenon, and this is supported by the cognitive sciences which have effectively reduced mind to memory/brain. In fact, a close examination of the concept of sixth sense, as it has been conceived of in the East, will demonstrate that the Buddhist and Vedanta sixth sense of mind is also a direct reference to memory. The only mystery about the sixth sense is how and why memory becomes so distorted that it creates a world of illusion for us humans.

In Eastern philosophy, it is argued that the sixth sense is caught up in a subtle confusion that blinds humans to their natural enlightened state. The main thrust of philosophical and religious endeavor in both Vedanta and Buddhism is working out the details of how mind (*citta*) has become so deeply confused. Philosopher and Buddhist David Loy said,

> "So Vedanta and Buddhism both emphasize the role of memory 'wrongly interpreted': identifying with such memories provides the illusion of continuity — a 'life history' — necessary to sustain a reified sense-of-self. Thus past and future originate and work together to obscure the present, usually negating it so successfully that we can hardly be said to experience it ..."[4]

Understanding the source of this confusion comes with the promise that the confusion can be overcome and the natural state of mind restored.

The Eastern concept of mind (*citta*) refers to the six consciousnesses (*vijñānas*). The first five types of consciousness are the sense of seeing, hearing, smelling, tasting, touching, and finally the sixth is the mental consciousness (*manovijñāna*). This Eastern sixth sense of the mental has often given the Western reader the feeling that there is something mysterious that is being left out—some untapped aspect of mind that will grant access to enlightenment, if only the secret can be revealed. Whatever the sixth sense is, calling it *mental* certainly captures the phenomena because mental is a general, all-encompassing term for psychic phenomena. Mental is such a broad category because the sensing of all phenomenal experiences are mental experiences of the mind. All the senses are sensations that are only present to mind. It is not an exaggeration to say that the mind constructs and then interprets the outside world from the five senses, which can only be known as mental impressions in the mind.[5] The fact is, the term *mental* is so general that it tells us nothing about the sixth sense that we didn't already know.

Perhaps explaining these same six senses from the Western perspective will provide a clearer understanding of memory/mind. For instance, it can be said that the brain receives its ancestral memory (or phylogenetic memory) through the DNA, which dictates the physical structure and the hardwired aspects of the ancient brain. This is how the five senses are wired to feed into the brain. Thus the brain, which serves as a store of experience, learns through the conditioning derived from the five bodily senses. Likewise, in the more recently evolved forebrain, the sixth sense, is filled with a sense of self along with the cultural indoctrination into which all humans are born. This demonstrates that all of mind is basically memory—the ancestral, hardwired memory and the programmed, cultural memory. Note that this is another manifestation of the 'schism of the senses' (mind-body, self-other, etc.) mentioned in chapter seven (see p. 54).

So for citta (mind/memory), the first five consciousnesses of mind are closely tied to ancestral memory, whereas the sixth consciousness is dominated by the cultural memory. Mind/memory is filled with the experience of all the senses, but the interpretive overlay that forms the world-view is a product of our cultural indoctrination. All is mind and all is memory. But the focus of the problem of mental confusion is narrowed when the focus is *memory*, rather than the more diffuse term *mind*.

The point of dwelling on the concepts of mind and memory is to explain the problem in a way that the modern human might understand. Because modern humans see the world through the lens of the modern world (the scientific spacetime paradigm), the best understanding will come from a close examination of that lens through which our thoughts are shaped by the modern scientific culture. We have learned through the physical sciences and life sciences a description of a world in space and time. And it is a powerful description. The West, and in particular the cognitive sciences, have a lot to say about mind and memory.

It is important to point out that the very first definition given for the English word *mind*, in most English dictionaries, including the *Oxford English Dictionary*, is *memory*. Although most Western interpreters render *citta* as *mind*, it should be recognized an equally likely, and perhaps better definition of *citta* would be *memory*.

Contemporary Western science has essentially reduced the mind to an epiphenomenon of the brain, and in cognitive theory the brain is functionally nothing but memory. As psychologist Anthony McIntosh said, "Memory is not the domain of particular systems in the brain, but of the brain as a whole. Memory is what the brain does."[6] This reduction of the mind to memory is a consequence of advances in neurology, cybernetics, information theory, and cognitive theory. To be sure, there are still many scientists and philosophers that are reluctant to reduce mind to the brain. There is still a significant number who maintain that there must be something more to mind that can never be completely reduced to physics, chemistry, and biology—this line of thought is sometimes characterized by a belief in 'emergent properties'. In other words, out of the complexity of the brain emerge phenomena that are irreducible to more fundamental elements. By definition, 'emergent properties' are inexplicable. The problem with this attitude is that it explains nothing and seems no different from a retreat into mystery. And this thinking has the added disadvantage of advocating a duality that separates self and other—mind is inexplicably kept separate from matter.

Time Sutra takes a different approach by utilizing the Mind-Brain-Memory reduction of contemporary cognitive science and combines this with insight into the sixth sense taken from Eastern philosophy. There are advantages to be gained by calling the sixth sense *memory*, since this is a distinct cognitive activity that has been carefully studied. When we understand the sixth sense as memory we can narrow the focus from the broad nebulous concept 'mind' to a known faculty that has been closely observed. Given the previous assertions of *Time Sutra* that *time* is composed of memory and that *self* is a construct made of 'time and memory', defining memory as the sixth sense provides a different, new perspective.

Buddhist and Vedic literature both insist that the trouble with humankind is that we are caught in a delusion of our own making and if we could only straighten out mind (*citta*) we would discover the bliss of reality. In particular it is the mysterious sixth sense of mind that needs straightening out. There is general agreement in the East that the wrongful thinking is wound up in self-ego and the temporal world, and it is necessary to eliminate both time and ego in order to attain the nirvanic state. This brings us to the point—if both the self-ego and time are false constructs of memory, as has been argued previously, then understanding that memory *is* the sixth sense might be the key insight into what has been covered over in mystery.

The Dalai Lama notes that a "number of people have observed that Buddhism is not a religion in the true sense of the word, but, more properly speaking, is a science of mind."[7] So it is surprising that Buddhism with its long history of thoroughly investigating mental phenomena has curiously little to say specifically about memory.[8] If indeed, as Casey said, "It is an inescapable fact about human existence that we are made of our memories ..."[9], then why did the Buddhists not give their full attention to memory? It has been suggested that the conscious de-emphasis of time in the East has created a cultural perspective that is nearly unbridgeable to Western understanding.[10] "It is generally believed that one reason why East and West will never meet is that the Indians had no history until Greek historians taught them ... and that the Chinese ... had no knowledge of the nature of time and developed no science of mechanics."[11] Howard Trivers remarks that

"India is generally regarded as the least historically minded of the great civilizations ..."[12] Hajime Nakamura also said of Indians that they "have not exerted themselves to grasp the concept of time quantitatively, and have never written historical books with accurate dates."[13]

Memory was certainly not central to Buddhist thought but neither was memory ignored. The Sanskrit term for memory is *smriti,* but the intended Buddhist meaning of *smriti* is most certainly far different from the Western concept of memory. Buddhist meditation master Chogyam Trungpa equates *smriti* with 'awareness of memory', "not in the sense of remembering the past but in the sense of recognizing the product of mindfulness."[14] Citing memory (*smriti*) as the recognition of mindfulness is important because mindfulness is a practice that is central to Buddhist meditation, and what Trungpa is saying is that the practice of mindfulness is an awareness that is identical to *smriti.* In other words, mindfulness is a meditation on memory, so in a way, mindfulness might be said to be the Buddhist practice of anamnesis. Mindfulness is remembering the present timeless being that has been forgotten.

In the end, *mind* and *memory*, when considered in an enlightened way, are synonyms. This is what the Japanese Buddhist scholar D. T. Suzuki indicates in the statement, "*Smriti* or *citta* or *vijñāna*. They are all used by Áshvagosa and other Buddhist authors as synonymous. *Smrti* literally means memory; *citta,* thought or mentation; and *vijñāna* is generally rendered by consciousness, though not very accurately."[15] Whether it is called *memory* or *mind*, for the true sixth sense to emerge we must engage in a meditation on *smriti* or *citta*. Said another way, manifestation of the true sixth sense entails contemplation of memory and mind—an anamnesis that remembers by bringing to mind what we have lost.

Chapter Nine

Mind-Body, Self-Other, and the Explanatory Gap

The phenomenon of human memory is, and always has been, perhaps
the most central problem of philosophy and psychology because it is
at the root of the mystery of human consciousness ...
—Noll and Turkington[1]

Mind-body dualism is as old as philosophy—in fact, it probably
coincides with the advent of human culture. Even the most primitive
cultures possess religion, and religion is almost universally
characterized by a belief that the soul or spirit is distinct from the body
and survives the death of the body. That which is believed to survive
earthly existence is a mind-like substance which retains some or all of
the mental attributes of the living human. As such, death is not final.
Certainly the afterlife would have no compelling value if all of memory
is obliterated by the death of the body.

René Descartes is considered the father of the modern mind-body
problem. He felt that the body was extended in space and the mind was
extended in time (eternity), that the body was mortal and the mind was
immortal. Descartes invented the spacetime coordinate system that
became a foundation of modern empirical science. The mind-body
duality became the mind-body problem because all of the phenomena of
space and time could be accounted for in the Cartesian coordinate
system—except for the psychic experiences of mind. The problem then
becomes how to give a complete description of reality that also includes
a description of our psychic phenomena. If you think about it, our

mental experience constitutes our only direct experience of the world. British philosopher Bertrand Russell (1872–1970) said it well, "What I maintain is that we *can* witness or observe what goes on in our heads, and that we cannot witness or observe anything else at all."[2] We think that we hear, see, and feel the outer world, but it is always the brain's interpretation of the external that we experience.

The mind-body duality manifests in a number of different ways. One common expression of the duality is the *self-other*; other variations are *subject-object, interior-exterior, mental-physical,* and *ideal-material.* All are expressions of the same underlying opposition that is formed when the sixth sense is split off from the other five senses. Immanuel Kant described it well by saying that the five senses of the body are a sense of space, whereas mind is the sense of time. For Kant even space and time are on opposite sides of the mind-body duality—thus *time-space* is just another expression of the same duality. Any philosophy or science that claims to describe reality must bring together all sensory phenomena into a unity, but Kant was never able to put it all together.

Toward the end of the 19th century, American philosopher and psychologist William James (1842–1910) laid the ground for the science of modern psychology with his publication of *Principles of Psychology.* He suggested a *Radical Empiricism*[3] by which psychology might reduce some aspects of mind to empirical observation, but he was adamant that a description of mental phenomena was necessary to any complete description of the world. German philosopher Edmund Husserl (1859–1938) recognized the importance of Kant's insight that the five senses are of the body and that time was a sense of the mind.[4] He proposed a *Phenomenology* that acknowledges mental phenomena must be included in any explanation of reality.[5] Husserl embarked on an analysis of psychic phenomena with the idea that *time* was the key to unraveling the psychic knot of time-ego-memory. The following from the *Cambridge Dictionary of Philosophy* is a good description of his insight: "The transcendental ego, the source of all intentional acts, is

66

constituted through time: it has its own identity, which is different from the identity of things or states of affairs. The identity of ego is built up through the flow of experiences and through memory and anticipation. One of Husserl's major contributions is his analysis of time consciousness and its relation to the identity of the self, a topic to which he often returns."[6] Husserl understood that there was an intimate relationship between time and memory, but he never resolved the nature of that relationship.

Late 20th century cognitive sciences focused on the connection between the two sides of the mind-matter dualistic opposition, and how a unification might be worked out. The lack of an explanation of the *mental,* in terms of the *physical,* has come to be called the *explanatory gap.*[7] Said in another way: the physical sciences (chemistry, physics, biology, etc.) to date have been lacking in their explanation of how mental phenomena are derived from a material world—this lack is the explanatory gap. This question is seen as absolutely central to resolving the mind-body problem.[8] No adequate answer has been forthcoming regarding the explanatory gap, but such an answer would profoundly change our worldview.[9] Explaining the 'gap' is tantamount to explaining consciousness, thus answering one of the most difficult and fundamental questions in science.

Several prospective solutions to close the "explanatory gap" have been put forward, of which the *materialist* argument has garnered the greatest support. Materialists maintain that the entire mind-body problem is ultimately reducible to physics, chemistry, and biology. Impressive advances have been made in understanding mental function at the cellular and molecular levels which suggest that ever increasingly detailed findings will eventually yield a complete scientific description of mind in physical terms, thereby unifying the duality. However, the materialists have failed to convince many in the philosophical and scientific communities that mind is fully explicable in material terms.

The counter-argument to the materialist reduction of mind-to-matter is that, regardless of how detailed the physical description, it can never satisfactorily explain how conscious experience arises out of physical and chemical reactions. Those who subscribe to this point of view are of two minds: those who accept dualism and believe that reality is composed of two incommensurable substances, mind and matter; and those who subscribe to *emergence* and believe that mind emerges out of the complex neural organization—how this emergent 'mind' occurs is simply beyond explanation—the effect of complexity is not reducible to more fundamental elements.

The debate over the mind-body divide has gone on as long as there has been philosophy, and it leaves many feeling pessimistic, thinking there is no possible answer. Philosopher Douglas Hofstadter said the problem is "as far from having an answer today (or, for that matter, at any time in the future) as it was many centuries ago."[10] Philosopher Ned Block said "The explanatory gap exists—and we cannot conceive how to close it—because we lack the scientific concepts. But future theory may provide those concepts."[11] Philosopher Michael Tye states, "It is just that with the concepts we have and the concepts we are capable of forming, we are cognitively closed to a full, bridging explanation by the very structure of our minds. There is such an explanation, but it is necessarily beyond our cognitive grasp."[12] Another suggestion by psychologist Jaak Panksepp is that "the explanatory gap is constructed by the ways we think about these matters linguistically ..."[13] Panksepp is partially correct, language does play a major role, but a purely linguistic explanation, which doesn't at present exist, would still be inadequate.

The explanatory gap, by its very name, is begging for an explanation, but the *emergence* argument is no explanation at all, and the *materialist* argument is incomplete in that science doesn't yet claim to know enough to give an adequate explanation. There is one other argument, overlooked by most debating the issue, and this argument actually does offer an explanation. American philosopher Daniel Dennett recognized where we should be looking—he recommends examining the software that is running on the brain if we want to understand consciousness.[14] The software that is most interesting, broadly speaking, is the software that comes to us through our cultural indoctrination and which prescribes, more or less, the way we think about the world. If we think about it carefully, it is aspects of our enculturation that open the explanatory gap at the heart of the mind-body problem, a problem that is central to our understanding of consciousness.[15] The mystery of the 'gap' and the mystery of 'time' is the story of a deeply buried mental confusion that distorts human perception of the world.

Dennett hypothesized there must have been a major transformation in the software of the human brain's operating system that propelled the cultural changes that took place a few thousand years ago. This is an adaptation of American psychologist Julian Jaynes' thesis of a transformation in the human mind in the distant past.[16] Dennett accepts almost none of the assumptions that Jaynes incorporated in his hypothesis, but does accept that some dramatic change in the software was necessary. Dennett reasons that "the underlying hardware of the brain is just the same now as it was thousands of years ago."[17] And he agrees with "Jaynes' idea is that for us to be the way we are now, there has to have been a revolution—almost certainly not an *organic* revolution, but a *software* revolution—in the organization of our information processing system, and that has to have come *after* language."[18] Dennett never specified what this software revolution entailed, and he never suspected that the software revolution was the invention of time itself.

The software revolution that profoundly influences human thought can be described in a variety of ways. Dennett expresses doubt about recovering knowledge of this 'software revolution' because it would require a 'software archaeology': the record is all but lost—cultural software is never preserved in a preliterate culture and can only be inferred from cultural artifacts. However, if the revolution was an 'upgrade' to the human brain's operating system, then that same software code is still presently with us—all that needs to be done is to uncover the current cultural instructions.[19]

Dennett's proposed 'software revolution' coincides with the Axial Age in the 1st millennium BCE (also called the Great Transformation),[20] but this is not a coincidence. An important thread of *Time Sutra* maintains that it is this historical period when a cultural change occurred in the interpretation of memory. The sixth sense, which always carried a subliminal feeling of the temporal, was transformed into the very conscious sense of time.[21] Time erupted into consciousness and human behavior came under the control of the ego.

As for the actual software that created this revolution, it is nothing more than an ingrained way of interpreting memory to be a representation of the past. This is a story that the child learns to tell about himself or herself, and when the story becomes sufficiently elaborate, the story becomes the self, and the storyteller takes on the persona of the adult ego. The story makes us into what we think we are, and we are made of those memories that stand as proof of the self. This is how the sixth sense becomes the *self* and all else becomes *other*. This is the origin of the mind-body problem.

Deeply inscribed in the brain is the encoded software that guides the interpretation of the experience of memory. The social indoctrination that directs this thinking begins in our infancy and this thinking follows us to our grave. We don't know the wiring diagram, but we can know that there are broad, permanently etched, neural pathways. We can't know how the code is written, but we can know that it effectively dictates the command that *memory is past*.

The explanatory gap is caused by a schism in our senses of the world. It is the segregation of the sixth sense from the other five senses which divides the world into self and other. Self stands on one side of

70

the gap looking across at the world on the other side of the gap. The division alienates the self from all of nature. Self becomes Ego by claiming mind and memory as its own and setting itself apart from the world.

The gap is a culturally induced software problem that is imprinted, if not globally throughout the brain, then at least operating at every point where the interpretation of autobiographical memory occurs. The answer to the explanatory gap is that the culprit software is a programmed instruction that specifies that *memory-is-past*. In other words, there is a viral meme hidden among the cultural beliefs, and it splits the world into Self-Other. The programmed message is simple, but the consequences are profound.

The explanatory gap is effectively explained, but the gap is a function of time, so the self-other schism will persist until the self is rendered timeless. We are brought back to a meditation on memory—*anamnesis* is the act of searching for a practice that restores the primordial sixth sense.

Chapter Ten

Reason, Sense, and Modernity

Can we remember to remember? Can amnesia give way to anamnesis?
—E. S. Casey [1]

Our experience of the world is often determined by our conditioned system of beliefs. The beliefs we hold to be the truest often lie buried, unexamined, yet controlling how we think and behave. Most humans pass their lives in a near somnambulant state—never seriously questioning their own belief system. This unquestioning allegiance to mental habits led Socrates to say, "The unexamined life is not worth living."[2] This is a harsh judgement against those who never bother to look. Even before Socrates, Heraclitus said, "What is not yet known, those blinded by bad faith, can never learn."[3] We are closed off from knowledge by our beliefs. If indeed a central problem of humans is attributable to the unrecognized beliefs that reside in memory, and which direct our behavior well below the conscious level, then it would be good to root out these unacknowledged beliefs.

A method of doubt is needed that not only allows the discovery of hidden beliefs, but grants the power to neutralize those beliefs. Earlier, in Chapter 7, it was pointed out that *belief* and *doubt* are opposite sides of a duality and that doubt cannot arise in the absence of belief. If there are no beliefs, then there is nothing to doubt. Wittgenstein said it well, "The child learns by believing the adult. Doubt comes after belief."[4] By invoking doubt we are not trying to achieve a world of *doubt*, but on the contrary, doubt itself is eliminated when we achieve the elimination of *belief.* And because there is nothing further to doubt, after we have doubted all the way to the end, our experience will become a reality that cannot be doubted.

The beginning of modern philosophy and the inception of the Modern Age is often dated to the works and times of René Descartes.[5] Descartes, commenting on his own unfounded beliefs, wrote, "Some years ago I was struck by the large number of falsehoods that I had accepted in my childhood, and by the highly doubtful nature of the whole edifice that I had subsequently based on them."[6] Recognizing the complete lack of ground on which his thought was based, Descartes set out to describe a method by which he could arrive at a foundation of complete certainty. He systematically proceeded to doubt all of his beliefs until he had distilled it down to what he felt to be unshakable ground. This technique of systematically questioning our assumptions has come to be known as the Cartesian Method or Cartesian Doubt.

From our contemporary perspective, advancing a methodology of 'doubting belief' hardly seems like a revolutionary idea. After all, a *belief* is at best a presupposition and at worst a logical error that is at odds with reality. Austrian born British philosopher Karl Popper stated the obvious, "Belief, of course, is never rational: it is rational to *suspend* belief ..."[7] However, in Descartes' world, the Thirty Years' War was raging and the killing was, in large part, over beliefs. Expressing the wrong belief was something that could get you killed, so Descartes was justifiably reticent in not pushing his view too openly. Actually, things in the present day haven't really changed all that much—the wars that raged over the globe in the 20th century, continuing until today, have been perpetuated by ideological beliefs about religion, economic theory, nationalism, etc. Still, in modern Western democracies, criticizing the accepted beliefs of the day is at least tolerated. Krishnamurti offered a categorical criticism of these accepted beliefs when he said, "It seems to me that all ideologies are utterly idiotic."[8]

René Descartes in Europe and Francis Bacon in Britain were both searching for reliable methods that would exclude nonsense from the natural sciences. Bacon advocated the necessity of empirical observation and experimentation, which, combined with Cartesian doubt, laid the ground for modern science. This is the marriage of sense with reason—the coupling of empirical data of the senses with a reasoned method for eliminating unfounded suppositions. Descartes' own recommendation on how this bias-free state of mind might be achieved was to "doubt, as far as possible, all things."[9] In the end, Descartes felt he had disposed of every possible belief and had achieved an indubitable ground on which to construct his natural science. He famously arrived at the conclusion, *cogito ergo sum*—I think, therefore I am.[10] He had validated his own existence, but with notable, unanticipated consequences. The beliefs that Descartes was unwilling to part with are beliefs that even now we still carry with us. He believed in a Soul/Ego and he believed in God. His Ego was in time and only God could save him from time. Upon close inspection these two beliefs form a kind of duality—there is no God unless there is someone who believes in God, and there is no transcendent Ego unless there is a God to grant everlasting life and other favors.[11] It has long been said 'man created God in his own image'.[12] By retaining the belief in God and Ego, Descartes forfeited the opportunity to move toward elimination of the mind-body problem. He did not create the problem, but he reinforced the gulf between body and mind.

Although Descartes mistakenly preserved the mind-body duality, he did make an important contribution to philosophy through his introduction of a powerful critical method, which if followed, would uncover and eliminate the spurious suppositions that infect our view of the world. The Industrial Revolution and all the technological innovations since the beginning of modernity are testimony to the incredible success that has been achieved through practicing this critical method. We would not have had modern chemistry and physics without it. But, over time, the fundamental questions that were most important to natural philosophers were eclipsed by the more mundane questions posed by those who had come to call themselves scientists. And these scientists were mostly concerned with engineering the natural environment. The big questions were turned over to the theoretical physicists/cosmologists—any natural philosopher still considering the big questions was pejoratively said to be indulging in metaphysics, which was not of particular concern for science. Science, for most practitioners, had become an ideology, specifically those scientists who hold the belief that the modern time-space paradigm is the reality. Modern scientists seldom remember the deep philosophical roots that are the actual source of their science. Distracted by, and satisfied with their success in manipulating nature, the philosophical ground was no longer a subject of intense investigation. Modernity remains, even today, unfinished.

Modern science picks apart every detail of the nervous system in the hope of unifying mind and matter, and yet the perplexity of the 'explanatory gap' stands between mind and body. Confused by this schism between self and other, and thinking that this schism is the nature of reality, we become convinced that we are an ego experiencing a journey in time and space. Ego, not knowing its own origin, then examines the seemingly inexplicable mind-body gap—a gap that is incomprehensible to ego, in part because the explanation denies the existence of ego. Ego is seeking an enduring meaning and purposeful existence, and strives to believe in eternity, becoming completely closed to the possibility of timelessness. Ego is striving for a salvation from time, not understanding that ego itself appears out of time—ego is a product of time. Like other viruses that elude detection by hiding, the ego is well protected by our doctrinal ways of thinking.

The ego, through the methodology of modernity, discovered a way of manipulating nature in time and space, with the great promise of making existence more comfortable and more meaningful. Four centuries later we still look toward the future for a better world. But anyone who has been paying attention is likely to see a grotesque future dystopia. Modernity seems to have granted us enormous power over nature, but this is coupled with a basic misunderstanding of our own nature. The postmodern movement recognized the failures of modernity, but being unwilling or unable to reason modernity through-to-the-end has meant that the profound teaching of modernity has gone unrealized.

Following the critical doubting method all of the way back to the origin reveals a truth about the world that heals the schism between humans and nature. The explanatory gap only appears to be an unbridgeable gap when there is a mental attitude that sets the sixth sense of memory into opposition to the five senses of body. The mental mistake is thinking that *memory is past*, which creates a schism in the world, a schism that converts reality into illusion. The method of doubt, followed without error, removes the mind-body duality. The explanatory gap has never existed except in the confused mind of ego.

Descartes set us out on the path of modernity—four hundred years later there is a solution, if you avail yourself of it, but to do so you must follow the methodology to its logical end, which leaves the self-ego staring into the abyss.

Part III

Finding Time

Chapter Eleven

Constructing Time—Time Reversal and Time Duality

For the universe, the two directions of time are indistinguishable ...
—Ludwig Boltzmann [1]

Time reversal[2] has been a topic of discussion in physics for as long as it has been recognized that the laws of physics are time symmetrical. Regardless of which way the arrow of time is directed, toward the future or toward the past, the laws of physics still equally apply.[3] Surprisingly, most physicists tend to agree that the human experience at any particular time will appear exactly the same whether time is running forward or backward. Physicist Paul Davies describes the experience:

"Our brain processes depend on the same physics as the rest of the universe, so they too would be reversed in a time-reversed world, along with the stream of consciousness and the memory and reasoning processes that attach to them. In other words, we would perceive and think in reverse in such a world. Our mental activity, including logical reasoning and concepts like causality and rationality, would all be inverted too. So a time-reversed being would not feel time-reversed at all. To it, all would appear normal."[4]

However, it is common for physicists who consider time reversal, and who acknowledge that a reversal would go unnoticed, to also persist in thinking that the two directions can be differentiated from each other. They claim the 'theory of entropy' is what allows them to attribute an 'arrow of time' to our existence.[5] The speculation of 'reversing entropy' is often a thought experiment associated with the Big Bang theory of an expanding universe. The universe that formed from the Big Bang is expanding and moving 'forward' in time, with ever increasing entropy, thus a collapsing universe would move 'backward' in time with ever decreasing entropy. But all of this speculation about entropy is futile as we saw in Chapter 4. Entropy is an illusion created out of a probabilistic model that assumes the motion of particles are random.[6] When analyzed, entropy theory will always predict a higher entropy than at the present, and in *both directions of time*, the past as well as the future. This again makes the point that time is identical in either direction, in which case there is no arrow of time.

The complete absence of any physical arrow of time has brought both philosophers and physicists to a crisis that they do not know how to confront. Physicist and philosopher of science Henry Mehlberg, who was obsessed with the problem of time reversal,[7] gives a clear explanation of the problem:

"... it seems obvious that, irrespective of any success achieved by so many sciences in getting rid of time's arrow, the very scope and perseverance of the efforts made in various fields with a view to re-establishing time's arrow deserves the philosopher's close attention and testifies to the seriousness of the underlying motivation. The nature of this motivation aiming at the safeguard of time's arrow is, perhaps, partly accounted for by the admittedly ever-shrinking basis of the scientific basis for temporal anisotropy.[8] The point at stake, however, is rather related to the fact that several deep-rooted beliefs concerning time, which came to be built into the very logic of ordinary discourse and to affect decisively the general outlook of mankind, seem incompatible with scientific findings concerning time ... That is why it comes as no surprise that the assumption of an arrowless, isotropic time is often viewed as a philosophical disaster..."[9]

Nevertheless, time appears convincingly real. How are we to explain this fact of the appearance of time when there is no empirical evidence for the existence of time? There is a consensus among some physicists and philosophers that if there is no physical evidence of time, then of necessity it must be purely psychological. Mehlberg himself stated that the temporal order is only derived from the psychological, "...if there are cases where the temporal order is directly and indubitably given, it is among our own psychological states that they must be sought, since these states are the only reality which we know directly and indubitably."[10] Australian philosopher Huw Price said that it is the way that we look at the world: "The perspectival solution to this problem is to say that the asymmetry of causation is a kind of projection of some internal asymmetry in us, rather than a real asymmetry in the world. In effect, the reason we see asymmetry everywhere we look is that we are always looking through an asymmetric lens."[11] The American particle physicist Victor Stenger, author of *Timeless Reality* said, "I feel that the possibility of time-reversal has been widely neglected for the wrong reason—a deep prejudice that time can only pass from past to future. Evidence for this cannot be found in physics. The only justification for a belief in directed time is human experience, and human experience once said the world was flat."[12] J. T. Fraser, who dedicated his life to the study of time said, "Let me return to 'one of the long-standing mysteries of physics: the origins of time's arrow' ... the 'arrow of time' is definitely not a problem that physics can solve. It is minimally a life phenomenon, and therefore its study belongs in biology, psychology, sociology, and the arts and letters."[13]

Physics and philosophy both point to the direction of time being psychological. There is no empirical evidence for the arrow of time, and without an arrow there is no time. Knowing that time is not real raises the most difficult question: How is it that we come to believe so thoroughly in something for which there is no rational or empirical evidence? M. F. Cleugh made this point when she said, "It is doubtless interesting to know that time is not real, but only apparent, but it is not sufficient. The denial of the reality to time avoids the main problem, for you have still to explain the *appearance* of time."[14] This is, at once, the toughest and the most important question to answer. The answer takes the form of an explanation of how we have deceived ourselves. The evidence cannot be found in the physical sciences, but elsewhere in the sciences there is a wealth of empirical data that gives all the details of this great deception in which we find ourselves. Physics denies reality to time, but it is the assembled arguments of psychology, information theory, neurology, anthropology, linguistics, etc. that actually explain the deception of mind. It comes to us as a complex cultural package, conveyed through the acquisition of our language.

Thus it becomes necessary to understand why our mind is so thoroughly convinced that time is real. Psychological studies demonstrate that time is learned (see Chapter 3), and from these studies it has been determined that it is learned through culture. Our culture is transmitted through our language, and culture and language frame the way we see the world. And it is this frame that we pass on to our offspring with the intent of perpetuating the culture. It is a fact that small bands of hunter-gatherers could only evolve into far more complex social units through the use of language, but this comes at the cost of becoming involved in the time construct. These primitive societies form a spectrum with regard to their involvement in time. Some, like the Umeda, have a very limited sense of time, but the terms they use can imply 'before and after,' which imparts a strong sense of temporal direction.[15] This type of temporal language undoubtedly infects the thinking of primitive societies, which respond through the use of ritual and ceremony to reconnect and remember timeless being.[16]

Universal throughout human cultures is teaching the respective ancestral lineage to the young and this teaching is perhaps one of the most ancient methods of the cultural use of memory. "History suggests that the oldest way of organizing information involved recalling one's ancestors, the line of descent that gave each person his or her identity as member of a tribe or a family."[17] To formally integrate into culture we learn our ancestral lineage and it naturally has an implication of 'not now' as regards the deceased ancestors. This is a teaching about something which is not now present, and if the ancestors are not now here, then when and where are they? The teachings are the present memories of the elders but the ancestors themselves are not present, and there are ancestors spoken of that not even the elders have a direct memory of. The ancient ancestors that are beyond any living memory belong to a mythical line of descent, which forms a vital element of cultural mythology.

Modern humans will immediately assume that the ancestral lineage is referring to the past, but this is not what primitive cultures intend when teaching the ancestral myths. Anthropologist Alfred Gell writes, "We have copious ethnographic testimony to the effect that various cultures do not consider that the temporal relationship of the present to the mythic/ancestral past is one that is effected by the passage of time: perhaps the best-known instance of this being the Australian Aborigines' beliefs about dream-time vs. the present."[18] In Jay Griffiths' study of time she writes, "For the Columbian Pira-parana Indians, the ancestral past, similarly, envelops the present. Rather than be cut off from it, the past is an alternative aspect of the present, approached through shamanism and ritual."[19] Historian, writer and philosopher Mircea Eliade explains, "This 'transcendent' world of the Gods, the Heroes, and the mythical Ancestors is accessible because archaic man does not accept the irreversibility of time."[20]

There is almost a logical force that compels us to view the generations of our ancestors, and know that the deceased ancestors are *past*, and *death* is inevitable, and that there is an unknown future. Then we become entranced with that future, because nothing is known about it, whereas the past is fixed, known, and done with. The whole fascination with the future is due to the fact that we *think* it can be made different because we *think* we have free will. This thinking launches us into the mundane existence of samsara, the *triple time*[21] of the past, present, and future. The fundamental assumption that promotes the fantasy of future is thinking that memory is representative of something that is not present, in this case the mythical ancestors. This is the assumption out of which time is created. But this in itself does not create time; time requires an arrow, and the arrow that is culturally assigned is determined by believing *memory is before*. It could just as well have been the alternative, *memory is after*, which would point the arrow of time in the opposite direction.[22] We believe in the former, yet recognize that the latter is absurd.

Why cultures construct time in the first place is attributable in part to the ancestral myth, but it is language that insidiously brings time to mind. First of all, language is the conduit for culture, so much so that it is often said, if the language is lost, the culture is lost. Language is infused with temporal thinking and thinking takes place through language, so it compels thinking temporally. Language is a serial organization of words, and the organization usually obtains its meaning from the serial arrangement. The order of the words of a sentence is time-like, in that they form a linear pattern, oriented in a particular direction, with a beginning and an ending. Language is deeply involved in time, and the speaker and the thinker, through engagement with language, become deeply involved in time.

Returning to the concept of time reversal, the only reason the issue of time reversal ever comes up is because we believe in the future-directed arrow of time. If we held no belief about memory, as opposed to the belief we all hold, which has been labeled the *Representative Theory of Memory*, then the idea of time reversal would never come up. But the fact that time reversal does come up, and has been widely discussed, reveals a lot about how we think about time. The central understanding to be gained from an examination of time reversal is that time is clearly fabricated out of the past-future duality. Humans tend to fall into the belief that the phenomena we experience as memory is 'time past'. This belief, whether it is derived from the structure of culture or from the biases of language, assigns an arrow to the direction of time, and so time appears out of the purely atemporal phenomena of our own memory. Keeping in mind that culture and language both exist in memory, it is hard to avoid the conclusion that our sense of time does not derive directly from our culture and language. Once memory is relegated to the 'time before now', it then becomes logically reasonable that there is also a 'time after now', which we think of as the *future*. The dualistic thinking that gives a structure to time becomes clear.

The past-future opposition, and the phenomenon out of which it appears is our own sixth sense, our own memory. Memory, even for humans, was originally timeless, just as it is for all other species. We have literally forgotten the nature of our memory. Culture, mostly through language, has changed the primordial memory into a temporal experience. The most damaging aspect of this alteration of memory is the outgrowth of the ego, which thinks that it knows, when in fact, it doesn't. From the ego comes the volition to act on the illusion of ego's perceived knowledge.

Memory is a fact of being, a fact of existence, it is a sense of the world, and it is drawn into time by the cultural teaching that proclaims memory represents *past time*, ignoring that memory is obviously an experience of now. Instead we pretend that it represents something that is 'not now'. What is not now is divided into past and future. It seems like a dualism that appears out of nowhere, but the *ground* is misunderstood memory. That which is made to seem like *before*, is a distortion of the primordial experience of the present sixth sense.

We have forgotten that we are already timeless beings. Ludwig Wittgenstein said, "If we take eternity to mean not infinite temporal duration but timelessness, then eternal life belongs to those who live in the present."[23] We have forgotten reality. We have forgotten who we are, because we have forgotten the nature of memory itself.

Chapter Twelve

Linear Time, Cyclic Time, and Rhythm

No one knows what time is; certainly no one knows how to define it and explain it to the general satisfaction. But we sure know how to measure it.
—Lewis Mumford [1]

Cyclic and linear time are the two dominant ways in which time is conceptualized. Understanding how these conceptualizations come about is easiest if we begin with the construction of linear time and then peel away the layers of accrued assumptions and beliefs that hide the underlying phenomena. Linear time covers over the more primitive cyclical time, which in turn obscures the primordial rhythm that is originally an aspect of the sixth sense (memory).

Linear Time: Modern humans believe time to be linear, which means that time moves from the already determined past, up through the present, and then flows on into the unknown future. Thinking in this way forms what we call the linear arrow of time. Linear time is indispensable to physics, and in general, necessary to all of modern science. The scientific requirement of time is that it be quantifiable. In other words, we must be able to measure time in some kind of equal units and ascribe number values to those units, according to their linear array.[2] The number values indicate where the unit of time occurs with respect to the other units along the linear array.

The number values that are assigned to the time units occur in numerical succession and form a linear series that science takes to be the time dimension. This has often been referred to as the spatialization of time because time is treated no differently numerically from any one of the three space coordinates. Taken together, the three dimensions of space and the one dimension of time form the paradigm of modern physics—the spacetime continuum. Albert Einstein explains how all of this is purely a conceptual invention of humans:

> "The psychological subjective feeling of time enables us to order our impressions, to state that one event precedes another. But to connect every instant of time with a number, by the use of a clock, to regard time as a one-dimensional continuum, is already an invention. So also are our concepts of Euclidean and non-Euclidean geometry, and our space understood as a three-dimensional continuum."[3]

Three questions come to mind. What are the units of time, how are they measured, and what is it that is being measured when we think we are measuring time? The units can be somewhat arbitrary, but are usually taken from some regularity that is observed in nature, and in this sense the units are grounded in actual phenomena. Four of the most accessible natural regularities that are adopted as units of time are the heartbeat (the second[4]), the diurnal period of light and dark (the day), the phases of the moon (the month), and the annual changes of the seasons (the year). These natural regularities are considered natural measures of time, but humans try to devise more exacting methods of measuring time. Skipping over the long history of time-keeping, we note that by the end of the middle ages and the beginning of the Renaissance, in European towns, clock towers were becoming increasingly common. These clocks were machines designed to mark the hours (an hourly reminder of time), and over time these machines were improved and miniaturized, to mark the minutes and eventually the seconds. Time consciousness is mostly due to being immersed in reminders about time.

Today we have incredibly accurate clocks that measure time, and they appear just about everywhere in our daily lives. The regular oscillation of a quartz crystal has been adapted to regulate the movement of very accurate and inexpensive clocks and watches. Any regular oscillation or rhythm can serve as a clock with which to measure time, whether it is the oscillation between day and night or the resonating cesium atom of an atomic clock.[5]

The final question: what is being measured when we say we are measuring time? What is this time that the clock claims to be measuring? It claims to be measuring the passage of time by the movement of the clock hands. The conventional clock has a sequentially numbered dial, and hands that swing around radially (clockwise) past the numbers. The hands represent passing through time, and the hands pass by numbers that represent progression along the time axis. The movement of the clock is intended to demonstrate the movement of time, but what exactly is being measured? The movements of the clock hands themselves are not time, but they are intended to represent the movement of time. Physicist David Park asks, "How quickly does time pass? At a rate of one second per second? That will get us nowhere."[6] The linear conception of time is a fiction, but it is constructed out of observable, naturally occurring rhythms, which also form the basis of cyclic time.

Cyclic Time: The time concept that is ascribed to most primitive[7] human cultures is cyclic time. Consequently, anthropologists and sociologists who have studied the problem of time have some important things to say about the nature of cyclic time. French sociologist Emile Durkheim (1858–1917) stated that time is actually a social construct that arises out of the social coordination necessary to human culture. "Durkheim argues that since all members of a society share a common temporal consciousness, time is a social category of thought, collective time is the sum of temporal procedures which interlock to form the cultural rhythm of a given society."[8] Durkheim's thoughts, published a century ago, hold a prominent place in the contemporary thinking of anthropologists. British social anthropologist Alfred Gell said, if Durkheim is correct " ... it suggests a way of resolving the feeling of perplexity which the notion of time itself has always seemed to generate ..."[9] In other words, time becomes much less of a mystery if it can be demonstrated that time is created out of the natural cooperation and coordination necessary to a cohesive social structure.

Durkheim thought that the notion of time arose out of the rhythm of our daily activities which, "In modern Western societies ... crucial elements of this framework are work and the clock. In primitive societies, however, temporal frameworks are arranged according to the 'periodic recurrence of rites, feasts, and public ceremonies."[10] Commonly it was thought that these recurring rituals and ceremonies impart a sense of cyclic time, but E. R. Leach takes issue with the idea that primitive cultures are experiencing a form of cyclic time. He argues that cyclic time is something that modern thought projects onto the actual primitive experience. "On the contrary, time is experienced as something discontinuous, a repetition of repeated reversal, a sequence of oscillations between polar opposites: night and day, winter and summer, drought and flood, age and youth, life and death. In such a scheme the past has no 'depth' to it, all past is equally past; it is simply the opposite of now."[11] When Leach speaks of 'past', it should be understood that he is actually speaking about 'memory', because a 'past' only arises from a mistaken assumption about memory being 'representative of past'. The past has no depth to it because the memory "of oscillation between polar opposites" in a primitive culture is not considered to mean 'before now'. Gell also tried to describe the 'primitive experience' when he said: "The flattened time is not even cyclical, it is simply alternating. The flow of time is like the flow of current in an AC electrical circuit."[12] Gell's explanation comes close but he can't get away from the idea of calling it time.

Leach goes on to say that the most elementary and primitive way of experiencing time is as a "discontinuity of repeated contrasts."[13] Leach is direct in saying that time does not exist except as a social construct, and he concludes by stating, "We talk of measuring time, as if time were a concrete thing waiting to be measured; but in fact we *create time* by creating intervals in social life. Until we have done this there is no time to be measured."[14] The fundamental phenomenon that remains after stripping away both linear time and cyclic time is the underlying *rhythm.*[15] The term 'rhythm' carries with it a time connotation, but used in the present context, it is a timeless rhythm.

Rhythm: By the late 19th century there was speculation that the sense of time was closely tied to rhythms in the world. Philosopher Howard Trivers in his perceptive study, *The Rhythm of Being: A Study of Temporality,* stated the following:

> "It has long been conjectured that rhythmic processes might be the basis for human time perception. [Ernst] Mach in 1885 suggested that 'the perception of time is closely related to processes repeating themselves in a periodic or rhythmic manner.' [William] James in 1890 and [Wilhelm] Wundt in 1911 are also said to have supported the concept of periodic processes as the basis of time perception. Thus psychologists have expressed a view corresponding to Einstein's, namely that periodic occurrences are psychologically as well as perceptually prior to time."[16]

It is instructive to look at how Albert Einstein grappled with these same issues of linear time, cyclic time, and rhythm. Most importantly, after a lifetime of considering temporality, he concluded that rhythm is more primitive than time. Einstein knew that time is learned, rather than innate, in part because Jean Piaget (with whom Einstein corresponded) clearly demonstrated that time is culturally acquired in childhood (see Chapter 3).

Einstein asks the important question, "But what about the psychological origin of the concept of time?"[17] He then answers, "Of itself it is doubtful whether the differentiation between sense experience and recollection (or a mere mental image) is something psychologically directly given to us."[18] Einstein is talking about the differentiation between the five physical senses and recollection *(memory).* Einstein fully realizes that memory is "considered" as *being* before now. He continues, "An experience is associated with a 'recollection', and it is considered as being 'earlier' in comparison with 'present experiences'. This is a conceptual ordering principle for recollected experiences, and the possibility of its accomplishment gives rise to the subjective concept of time, i.e., that concept of time which refers to the arrangement of the experience of the individual."[19] Einstein not only has a firm grasp of how the "subjective concept of time" arises, but he also understands that it forms the temporal "arrangement of the experience of the individual". Einstein has sketched an outline of how the self-ego is conjured out of time, due to our thinking that memory is "earlier".

Howard Trivers, in *The Rhythm Of Being: A Study of Temporality*, writes a great summary paragraph about Einstein's temporal conception:

> "A 'periodic recurrence' is a reiteration, and as such, it is also a rhythm. When Einstein puts the concept of periodic occurrence ahead of the concept of time, he is testifying that rhythm is the primordial ground of time. He is also testifying that the common view of periodic recurrence or rhythm as necessarily presupposing time is not correct in a fundamental sense. To be sure, once the concept of objective time has been elaborated, we can interpret rhythms in a temporal way, and we can measure their temporal intervals. However, their common usage neither reveals nor corresponds with the fundamental conceptual relation."[20]

Einstein has shown his deep insight into how time was constructed.

Conclusion: Much effort was expended by Edmund Husserl, Martin Heidegger and others attempting to decipher the phenomenology of time, and generally, getting it mostly wrong. However, using the above arguments from anthropology and physics, it can be stated that the phenomenon of time can be reduced down to two essential components, rhythm and memory. And rhythm only has a time-like quality when memory is presumed to be past. So, there is only one true phenomenon responsible for experience of time and that is the phenomenon of memory—time occurs when infected by the meme, *memory is past*. It is important to recognize the phenomena of rhythm, but it is critical to understand that rhythm is perceived through memory, and if memory is infected with the temporal meme, then rhythm takes on a time-like character.

Rhythm, as has been explained, is an oscillation between two opposite poles. We can only recognize the major life rhythms through memory. We all live by the natural rhythms, and even the modern human feels the fundamental rhythm of night and day. We know the rhythm. When it is light we know the dark, and when it is night we remember the day. It is through memory that we know the diurnal rhythm. When it is summer we remember the winter, and in the depths of winter summer comes to mind. This is memory that allows us to recognize the annual rhythm. When we think of these as cycles we are temporalizing the rhythms into a circular time with an arrow. Many anthropologists have realized this error and have described the rhythms as oscillations between two poles of difference. The oscillations are not thought to be in time, rather, they are the natural rhythms of the world.

Clock faces turn the diurnal rhythm of life into successive numbers, and as the hands turn twice through the circle of numbers, the diurnal cycle is repeated and we use a calendar to count the repetitions, thus our days become numbered, and these are further organized into the cycle of seasons so that we can number the years. But the cycles of a clock dial don't make time cyclical, nor does linking up the cycles together make time linear. Clocks do help us conceptualize both cyclic and linear time, though both are constructions of our imagination.

Mainstream anthropology has determined that many primitive societies live in near timelessness. The investigations of the psychology of child development demonstrate that time is a belief that is taught by the culture in which the child is raised. Since early in the 20th century physicists have increasingly come to think that time is *not* a fundamental aspect of reality. And back at the very beginning of philosophy, there was doubt about the reality of time. After all, the first truly systematic philosopher, Parmenides, concluded that reality was timeless.

All the religious ceremonies and all the religious rituals of a culture are signs of the fallen human—the human that has fallen into time. All the religions and all the philosophies are a sign of humans' fall into time, because, universally, these are attempts to extricate the self from the consequences of time. But almost none of it works because there is a lack of understanding of how time works. Ironically, most of these efforts are made on behalf of the ego, but the ego itself is only a fiction that arises out of, and then perpetuates, the temporal illusion. The illusion is far more than a simple illusion, it is a mental virus that compels us to think and act the way we do.[21] We are no more aware of these hidden motivations than the ant climbing to the tip of a leaf of grass, positioning itself so that it is more likely to be eaten by the cow, in order that the parasite that has programmed the ant's behavior can complete the next stage of its life cycle in the grazing animal's liver.[22] For humans, our brain has been programmed by a cyber-virus and we don't know it. We have no clue as to what the program is or how we were indoctrinated into following the program. In the end, we are subjected to the consequences of time, not knowing that all the trouble time brings is something which might be avoided. Eckhart Tolle, in *The Power of Now*, explains that our troubles are derived from a fixation with past and future:

> "All negativity is caused by an accumulation of psychological time and denial of the present. Unease, anxiety, tension, stress, worry—all forms of fear—are caused by too much future, and not enough presence. Guilt, regret, resentment, grievances, sadness, bitterness, and all forms of non-forgiveness are caused by too much past, and not enough presence."[23]

Chapter Thirteen

Time-Space Duality and the Block Universe

Not recognizing that we are dreaming, there is no way
to imagine awakening from that dream.
—Dzogchen Ponlop [1]

The following is a psychological description of how we think and talk about time and space. First, employing the language of oppositions, an explanation of the time-space duality will demonstrate how this complex notion influences our thinking. Then, a discussion of how modern physics explains the same time-space duality as it is conceptualized in the terms of the block universe.

Time-Space Duality: The time concept is entangled in two dualities. Time arises out of the first duality through the simple assumption that *memory is past*. But then this fabricated time-concept enters into a mutual relationship with space, which forms the second duality. The first duality, the past-present-future concept, originates from assuming *memory is past,* and gives time its direction. This first duality (call it the *past-future* duality) has been explained in some detail in chapter 11. This is the basic substrate on which the second duality is built. The second duality is the *time-space* duality. Time-space is like most other dualities: both sides of the duality are dependent for their existence on the presence of the opposing side. Whether *hot-cold, good-bad,* or *future-past*, if one side of the duality doesn't exist, the other side no longer has meaning. Dualities are always self-referential and they cannot exist without their opposing referent. Quite simply, a word that refers to nothing is devoid of meaning.

The time-space duality, however, is in one way different from most other dualities in that it is not an opposition of two terms, rather the terms cooperate. Both terms (time and space) depend upon each other in support of the spacetime continuum. Each concept is necessary to the spacetime continuum if there is to be *change* or *motion*. It is hard to imagine what change would be when there is no *space* in which to move, and it is hard to conceive of movement, without a *time* in which motion occurs. Time and space are both equally necessary to the narrative of an ego existing and acting in space and time.

The past-future duality, as previously discussed in chapter 11, is formed out of two *opposing* assumptions (unlike the *complementary* assumptions of the space-time duality). *Memory is past* is an assumption that is adopted by culture, and gives life to the concept of time by providing the necessary arrow of time. But, like most assumptions, it could have been assumed to be otherwise. Although culture has assumed that *memory is past*, culture could have just as well assumed that *memory is future.* Of course, making the assumption that memory 'represents the future' immediately appears completely absurd, whereas, it is with great difficulty that we come to an understanding that our existing assumption, *memory is past,* is equally absurd. This brings us to a point that Jacques Derrida makes—when something is stated, look closely and you can also find what is not being said. In this case the path not taken is not any more absurd than the path that we are on. The argument that *Time Sutra* makes is that the correct path would be to avoid dualism altogether.

Time and space, as they function in the time-space duality, are both closely associated with memory; in particular we assign a time and place to the autobiographical memories. We make the assumption that our present autobiographical memory, experienced in the 'here-now', is actually representative of elsewhere in *space*, and at some other *time* in the past. The 'place-and-date' stamped on autobiographical memory is the complementary space-time duality through which we interpret memory. In other words, the autobiographical memory comes with a tag of 'time and space' that means 'when and where.' This is the narrative of 'when and where' that tells the personal history of the ego. Our lives become the life of an ego that persists in space and time.

The ego clings to the great promise of modernity, the promise that the future will be, or can be made, better. The promise is that this *dis-ease* that we feel in the present will be alleviated in the future through the acquisition of something we don't currently possess. The promise of the future is why we strive in the present to obtain something that we imagine will exist in a better future. The present will always lack something that the future can provide—modernity is characterized by 'living for the future'. The present, for the temporally oriented self-ego, is only a means to an end, but at the end of ego's time, when the promise of future runs out, ego dies. The promise of future is also the promise of death. The temporal-ego is our troubled soul, which becomes the stimulus of all philosophies and religions; basically, these are all attempts to escape from the consequences of time. The inevitable death of self-ego in time, focuses the mind.

Humans are so good at believing in things for which there is no evidence—they invent childish salvations and then proceed to believe in them.[2] Philosophy is a systematic attempt, when done properly, to overcome naive thinking and realize the obvious, that reality does not require belief. Philosophy is a search to root out the beliefs that are concealed by the attitude of certainty. Philosophy asks: What is it when memory is the present? This is the *koan*[3] that forms the central meditation of *Time Sutra*.

The time-space duality confers a sense of free will on the self-ego, and this goes a long way toward making us think the way we do. Generally, throughout *Time Sutra,* the *time* aspect of the time-space duality is emphasized at the expense of the *space* aspect. But, the two form a true duality, so that each is dependent upon the other. However, there is a reason that the primary focus has been on the analysis of time rather than space. As Kant pointed out, the five bodily senses are of space, and since we live in a material-empirical culture, we take the external world (space) as reality. And, our spacetime paradigm, which is a reflection of the body-mind duality, does a wonderful job of measuring and analyzing the five objective bodily senses, but it does a poor job of analyzing the subjective sixth sense. Notice that the sixth sense only becomes subjective because the self-ego identifies itself with memory—it has claimed the sixth sense as its own subject.

Our five bodily senses are readily available to spatial-temporal measurement, and observations from these five senses, taken together, constitute the data that drives the empirical sciences. Time, however, is mysteriously different. Time is of mind—it is subjectively experienced, so we use external natural rhythms (clocks) as a substitute for our lack of objective experience. And because it is the sixth sense that is our sense of time, it thus seems more than justified, it becomes imperative, to focus on understanding the sixth sense—besides, we have already studied the phenomenal observations of the other five senses ad nauseam. Modernity has closely observed and closely measured the phenomena perceived through the five senses—and the data compiled, along with articles and books written describing and analyzing these observations, fill university libraries all over the earth. But it is the sixth sense of time, or more correctly, the sixth sense of memory, that holds all the answers to the most important questions.

Modern science tries to close the 'explanatory gap' through the multi-disciplined inquiry of the cognitive sciences. However, the gap can only be eliminated through the re-integration of the senses. This requires comprehension of the true nature of the sixth sense. The best metaphor for the sixth sense is *memory*—memory that is not mistaken for past.

Certainly the modern digital computers in many ways resemble our own minds because they are designed by minds to perform certain functions of the mind. It should be no surprise that there are close functional parallels. A central indispensable structure of a computer is the memory, just as the central physical structure of the brain is that of a memory. Whatever the egoic-self is, one certainty is that self-ego resides in memory because it is nothing other than what is remembered. In fact, it is not an exaggeration to say the ego considers all of memory to be identical with self-ego. This is the source of thinking that salvation consists of the survival of some aspect of the remembered self.

Humans have devised two common salvations, reincarnation and heaven. Reincarnation generally assumes that some aspect of our past lives is preserved and plays a key role in the present, but reincarnation usually offers some ultimate redemption that is heaven-like in its eternity/timelessness. Heaven assumes that the self-ego gets to persist forever, residing in happiness (unless, of course, you don't believe the myth, then you are condemned to eternal unhappiness). All that these childish salvations require is belief. Redemption or salvation ultimately, by this way of thinking, requires a belief that there is some preservation of the sixth sense of memory in *time everlasting.*

Time and space thinking are coequal in making a supposition about memory. What has been drilled into memory by our culture is not just *memory is past,* but also *memory is not here-now.* Jean Piaget was the first to document that time is a learned phenomenon (see Chapter 3), but he went on to also document that learning to think in the abstract terms of projection in space is learned, and it is learned at about the same period of child development that the time concept is learned. Piaget said, "Real projection begins to be grasped (after 7-8 years) ..."[4] Prior to Piaget, Bertrand Russell wrote in 1921, "It would seem that, in man, all that makes up space perception, including the correlation of sight and touch and so on, is almost entirely acquired."[5] Thus, memory is not only expanded into the temporal dimension by the assumption of the Representative Theory of Memory, but the Representative Theory also implicitly expands memory into space. Our culture teaches us to order our memories in time and space, thus forming a time-line that represents our personal history. We then feel like we exist in time, because we are constituted out of the content of our memory (we are, no doubt, what we remember), which we assume to be our personal historical past. The self-ego is, in fact, constructed out of nothing but memory, and the program that culture lodges in our brain organizes memory into a fabricated time-space duality, which obscures our own timeless nature. To glimpse the unadulterated phenomena, it is good to understand the nature of the delusion in which we are immersed.

The Block Universe: The foregoing is an investigation into the time-space duality and the consequent psychological state that makes us think the way we do. However, there is also a good theoretical description of the time-space duality in physics that is closely compatible with this psychological explanation. Physics, for over a hundred years, has entertained a model that has come to be called the *Block Universe Theory*. The American philosopher and psychologist William James was the first to use the term Block Universe.[6] The Block Universe is also sometimes called the Minkowski Universe, or Minkowski Spacetime. It was Hermann Minkowski, a former teacher of Einstein, who first developed the model, applying Einstein's *special theory of relativity* to the interpretation of the four-dimensional spacetime continuum. Howard Trivers explains that, "Minkowski's world is like the 'block universe', a term of William James, where events do not happen but we merely become aware of them ... According to this view, the passage of time is a feature of consciousness that has no objective counterpart."[7] J. R. Lucas describes it thus:

> "We picture worldlines in Minkowski spacetime as already timelessly existing and ourselves as going along a worldline encountering our predetermined future as it takes place—in Weyl's words, 'The objective world simply *is*, it does not *happen*. Only to the gaze of my consciousness, crawling upward along the lifeline of my body, does a section of the world come to life as a fleeting image in space which continually changes in time'."[8]

The Block Universe is usually graphically depicted as a sketch of a three dimensional block, made up of an infinite number of time-slices (instances). The time slices can be visualized like flat-lying sheets stacked vertically and arranged serially (from the bottom up), representing time from the beginning to the end. Think of a three-dimensional cube whose sides represent the boundary of all of space and time. The three-dimensional sketch shows two horizontal dimensions of space (x, y axes) while the vertical third dimension (z axis) is made up of time-slices (we are limited in our ability to represent a four-dimensional concept within a three-dimensional sketch, drawn on a two-dimensional piece of paper). The Block Universe description of our personal, individual experience is a time-line, often referred to as our *worldline* or *lifeline,* that passes through the four dimensions and imparts the sense of time. But those who look closely see that nothing moves or changes, everything is predetermined from beginning to end, and the physics implies that it can't be otherwise. We are moving along our lifeline and it seems like time, but physics informs us that this is an illusion.

The greatest physicist of this past century, Albert Einstein, along with who is arguably the greatest mathematician of that century, Kurt Gödel, see the world similarly to the way Parmenides saw the world 2500 years earlier. Gödel knew that the time dimension was no different from the three space dimensions. "Gödel would conclude ... that *t*, the temporal component of spacetime, was in fact another space dimension—not time as we understand it in ordinary experience."[9] And so the fourth dimension was left with no arrow of direction, but if we think as if it has an arrow, it makes it seem like we are in time. For the physicist, describing our existence in terms of the Block Universe, we follow our individual lifeline through space and time, which is physics' way of explaining why we experience a feeling of time. Complementary to this is the psychological thinking that the experience of memory is 'not now' and 'not here' (the time-space duality), and it is this thinking that forms the psychological explanation of why we experience time.

Karl Popper, in his autobiographical *Unended Quest*, writes that "I tried to persuade him [Einstein] to give up his determinism, which amounted to a view that the world was a four-dimensional Parmenidean block universe in which change was a human illusion, or very nearly so."[10] Dan Falk, in *In Search of Time*, said of Einstein's theory, "With special relativity, it seems that the block universe now has the backing of the greatest physicist of the twentieth century." Falk then asks, "What did Einstein make of this idea? His writing suggests that, much like Parmenides, Augustine, and McTaggart, he viewed the idea of time—or at least the "flow" of time—as something that resides not "out there" in the universe but rather within each of us."[11]

If we are in time, we become a *present,* passing along a lifeline in the time and space of the Minkowski block universe. Lost, we have the existential dread that one possesses when there is no possibility of meaning, because everything is fixed for all of time. Maurice Nicoll said it best, "Man has fallen asleep in matter and in time and in himself."[12]

Part IV

Thinking About Time: Three Philosophers

Chapter Fourteen

Derrida and Deconstruction

"At any rate it is now quite clear that neither future nor past actually exists."
—St. Augustine[1]

Introduction: *Time Sutra* is adamant that the genuine experience of reality is inaccessible to those who have not understood time. In other words, if you don't get time right, much of what follows will be distorted. Time is the central question of philosophy. Many philosophers who studied time have been mentioned throughout this writing, but four philosophers are given a closer scrutiny: Parmenides, Derrida, Nietzsche, and Longchenpa. Most of the major players in philosophy have developed some thoughts on time, and many have at some point thought long and hard about what time is and how it works. Throughout the text of *Time Sutra*, dozens of these philosophers have been mentioned or quoted. But now we will take a closer look at three of these four philosophers.

Parmenides' philosophy was treated in an earlier chapter. Because of Parmenides' importance to the Western tradition, it was necessary to introduce his thoughts early in the text. The present chapter, and the two that follow, will examine the philosophical thinking of the other three philosophers mentioned—each of whom considered time pivotal to their understanding of the world.

Up first is Jacques Derrida. He is chosen as a representative of the linguistic movement in continental rationalism, and as a postmodernist who was influenced by the mid-20th century existentialist philosophers. Second is Friedrich Nietzsche. He is discussed because of his influence as the most popular modern philosopher and he also makes an interesting story. Then, lastly, the great Tibetan philosopher Longchenpa is included to partially compensate for the emphasis on the Western tradition.

These three individuals have thought deeply about time, and each came to a very different conclusion. This narrow selection includes no modern empirical philosopher, but the modern empiricists are well represented elsewhere in *Time Sutra*. This three-chapter section provides a flavor of some of the many divergent ways in which time might be considered. Besides, *thinking about time* is a worthwhile practice for its own sake—this is what *Time Sutra* is all about. You can never wrap your head around time because it is illogical, but amazingly you *can* get your head around the illogic that creates time.

The first two philosophers, Derrida and Nietzsche, get time wrong but their analysis of time is well documented in their writing, and this allows us to examine how they reason their way into their beliefs. The other two, Parmenides and Longchenpa, get it right. The philosophy of Parmenides, the first recorded Western philosopher to correctly analyze time, was discussed in Chapter 5. The Longchenpa chapter will tell the story of *Time Sutra* from the view of a culture distant in time and space.

In the development of modern thought there are two principal competing ideas about what constitutes the source of knowledge. These two contrary sources of knowledge are *sense* and *reason*. When expressed as Western philosophical systems, the two are referred to as *empiricism* and *rationalism* respectively. Simply put, strict *empiricism* is the hypothesis that all knowledge comes through the senses, whereas radical *rationalism* entails thinking that 'true knowledge' can only be derived through reason alone. Both views about the source of knowledge coexisted throughout the development of modern thought, and there were times when efforts were made to reconcile both views into a single theory of knowledge.[2] Francis Bacon is usually identified as one of the earliest advocates of empiricism, whereas René Descartes represented the view of early modern rationalism, but elements of both were always present in the modern paradigm.

108

So how would these two competing philosophies see the world differently? The *rationalist* starts from the position that our sensory experiences of phenomena are inherently flawed, and that truth can only be found in *reason*. The rationalist recognizes the five senses of the body are imperfect, and concludes it is only mind (the sixth sense) where truth is possible. On the contrary, the *empiricist* understands that thinking is generally confused and cannot be trusted. For the empiricist the five senses constitute the total information available to us about the world. So if truth is to be found, it will be found in careful observation of what the five senses reveal.

Both philosophical approaches are prescient in their detecting that the various six senses are often distorted. However, they differ in their assessment of where the distortion lies, and this difference is one more expression of the mind/body duality. When the six senses are split between the five bodily senses and the sense of mind, we begin to think in a dualistic fashion. Rationalist and empiricist styles of thinking don't create this dualistic mind-set, they are just another expression of the consequence of believing *memory is past*.

Empiricism was the more prominent way of thinking throughout the British Isles, and so it is sometimes called British empiricism. Meanwhile, rationalism was more prominent on the European mainland, and for that reason it is sometimes called continental philosophy, or continental rationalism. Philosophically, this has been a recognizable schism for a long time, but it became a well-defined and divisive open dispute in the 20th century. Early in the century it was becoming clear that empiricism had dominated, if not outright won the debate, because it was the influence of technology and empirical science that made the modern world. At the same time, the rationalism of the continental philosophers seemed to be caught up in endless words and arcane disputes that only their ardent followers cared about.[3] At the conclusion of his best-selling book, *A Brief History of Time,* theoretical physicist Stephen Hawking mocked the state of philosophical rationalism:

"Philosophers reduced the scope of their inquiries so much that Wittgenstein, the most famous philosopher of this century, said, 'The sole remaining task for philosophy is the analysis of language.' What a comedown from the great tradition of philosophy from Aristotle to Kant!"[4]

Empiricism had won the day; scientists claimed the territory of empirical philosophy as their own and deemed that all else was 'metaphysics' and not to be taken seriously. Science had become the modern paradigm, and, more and more, science had exerted control over nature. The completely deterministic character of science seems to eliminate the possibility of transcendent meaning, or purpose in life. The modern paradigm offered a world that is a predestined, mechanistic existence that precludes human free will. How can any meaningful reality be salvaged from this state of affairs?

Early in the twentieth century, citizens of Western developed nations were seduced by an ever-increasing array of manufactured goods. By the end of the 20th century, the governments, corporations, and politicians referred to the citizens as consumers, and the citizens weren't offended because they had come to act like, and think of themselves as, consumers—grazing on the abundance of novelty, but always looking for something new. The first decade of this new century has been characterized by an endless stream of ever more entertaining products which have completely captured the attention of consumers. Present-day consumers are immersed in a cacophony of advertising and entertainment that can easily fill every waking moment, and the consumer gadgets most in demand are those that provide access to this frenetic stimulation. Should any tedium, despair, or depression slip through the onslaught of distractions, there is an abundance of pharmaceuticals to address the discomfort. The social climate has engendered an intolerance of both solitude and quietude that borders on the pathological.[5]

Trepidation about the dangers of technology is as old as Greek mythology,[6] but modernity has given us a greater magnitude of worry. By the mid-twentieth century, those who were concerned about the encroachment of technology had developed a severe anxiety derived from the growing sense of alienation from their own culture, coupled with an attendant need to be closer to nature. Existentialism[7] was an attempt to address this angst, and an insistence that we give meaning to our own existence. Existentialism was driven by a longing to be more in alignment with a nature closer to our own nature, and to again understand our own being.[8] Existentialism reached its peak influence shortly after mid-century. It had elegantly defined the problem, but it faded away without providing the much-desired solution, having never moved beyond the questions that existentialism posed.[9] The existentialist philosophers were very much concerned about time, but none of them realized that what had been lost was our primordial connection to timeless being. Our direct connection to nature is severed when we begin to think *memory-is-past*.

Derrida: Jacques Derrida (1930–2004) acknowledges his indebtedness to two of the most prominent existentialist philosophers, Martin Heidegger and Jean-Paul Sartre.[10] What is most important to know is where Derrida got his ideas about time and the nature of his philosophy of time. Although Sartre recognized time to be unreal, and knew that temporal thinking strongly influenced our behavior, time was not central to his philosophy and his ideas on time had little effect on Derrida's thinking.[11] It was Edmund Husserl's and Martin Heidegger's thoughts on the problem of time that exerted the greatest influence on Derrida's theory of time.[12] Derrida wanted to finish the analysis of time that was the focus of both Husserl's phenomenology and Heidegger's existentialism, because neither had, in Derrida's mind, developed an adequate theory of time. Derrida proceeded to describe a theory of temporality that was quite original but very flawed. However, his explanation of how he came to think about *time* is informative, and it contains sufficient detail to allow an understanding of how he went wrong.

111

It is difficult to follow some of the philosophical moves that Derrida makes. He was so notoriously hard to read that it prompted an exasperated E. O. Wilson to write, "Nor is it certain from Derrida's ornately obscurantist prose that he himself knows what he means."[13] But a careful reading reveals that Derrida wasn't trying to be difficult, only that the path he took could never provide more than a muddling explanation. Underlying all of Derrida's thinking is the implicit assumption of linguistic analysis which claims that analyzing how we talk and think about the world will tell us a great deal about the world. Those, like Derrida, who take the radical 'linguistic turn',[14] are thinking in the spirit of the continental rationalist, and as just noted Stephen Hawking said the rationalist philosopher thinks that "The sole remaining task for philosophy is the analysis of language." Rationalists, those who assume that ultimate truth is derived from reason alone, have a deep prejudice against the five bodily senses and this attitude precludes having an interest in closely examining the detailed world of empirical observations. So the rationalist will completely miss out on the empirical understanding that there is no physical basis to a sense of time. Time is something that is experienced only by the sixth sense, which is the experience of memory (mind), as seen through the distorting assumption that *memory-is-past*. Thus it can be argued that the rationalist's own thought and language is caught up in the corruption of mind. It is our thinking that splits the mind from the five senses, and in doing so, creates one of the most pervasive and intractable problems in all of philosophy, the mind-body duality.

There are three essential points that can be made about Derrida's analysis which explains how he arrived at his conclusions about time. First, Derrida uses the linguistic analysis of Saussure, who recognized that all language is self-referential, and all words express a difference.[15] Secondly, he notices the dualistic nature of language: for a word to have meaning there must be a word, or meaning, that opposes it. Derrida referred to these dualities as 'binary oppositions'.[16] Finally, Derrida reasons from this that when something is said it implies indirectly that which was not said.[17] He felt that understanding 'what was not being said,' would in turn lead to a deconstruction of what was being said. His analysis is somewhat correct. He thought (like Heidegger) that what was *hidden,* was hidden in the way we use language and was accompanied by a general misunderstanding of time. But when Derrida applies this analysis to the conceptualization of time, his analysis becomes confused and ultimately fails. He actually seems to reach the conclusion that it is the present that is unreal. "Derrida's principal thesis is a denial of 'presence'."[18] His philosophical reasoning is a case study in how an analysis of time can go wrong, and it demonstrates that when the puzzle of time is not solved correctly, everything that follows may be just an illusion.

In fairness to Derrida, the above quote, "Derrida's principle thesis is a denial of 'presence'," is somewhat over the top. What Derrida seems to be saying is that we over-emphasize the *present* at the expense of the *past* and *future*. He feels that if we can balance the prejudice for the present with the past and future, then we are closer to the truth. Derrida said, "The concept of time belongs entirely to metaphysics and it designates the domination of presence."[19] He is correct in saying that "the concept of time belongs entirely to metaphysics", but there is an actual phenomenon that underlies time (and Derrida is unaware of this). In other words, it is the sixth sense of memory (coupled with the assumption that the memory is before now) which supplies the feeling that time is *something*. But, in reality memory is not an experience of time, because memory is not past, rather, it is always an experience of the present.

Deconstruction: If any term of Derrida's philosophy is recognizable to a larger audience, it would be the term *deconstruction*. Derrida's philosophy is closely associated with the concept of 'deconstruction', a term he borrowed from Heidegger's use of *'destruktion'*.[20] Deconstruction means, as the word suggests, a 'taking apart' or 'disassembling'. However, for Derrida it is a linguistic maneuver that is employed to accomplish the deconstruction. And when he turns his attention to the analysis of time, it is the linguistic technique of deconstruction that he uses.

Derrida's method of deconstruction is a literary device that is a method of closely reading and carefully interpreting the text. Derrida wants to read into the text what is explicitly silent and opposite to what is actually said. He is correct that *presence* is caught up in an opposition, and as a consequence mis-perceived and misunderstood. But he makes a mistake in thinking that *presence* is one limb of the duality of opposition, whereas it is simply caught up in the dualistic opposition of past and future. *Past is the postulate*, and so if the postulate is accepted, then it is reasonable to believe in *future*, but this leaves *presence* straddled by this *past-future* duality, and yet *presence* is the only empirical aspect of reality. Before and after are suppositions—only presence is experienced.

Derrida felt that he should read the text for what was not being said (this is deconstructing the meaning of the text), and this sort of reading would divulge what was hidden and so provide the full context to what was actually being said. He then turned this method of linguistic analysis toward the analysis of time. We speak of the past, present, or future, but it is always the present that we give preference to,[21] thus Derrida reasoned that the past and future were of equal stature with the present, but generally left unsaid. From this he arrived at the idea that the error in our experience of reality comes from giving priority to the *present* at the expense of *past* and *future*.

Derrida misinterprets the past-future duality as an opposition to presence. He never saw that the dualistic opposition was actually between the *past* and the *future*, and he never recognized that the *present* was timeless. The present only gets caught up in the duality of time because the ultimate phenomenon of time, which is memory, is assumed to represent 'before now'. Thus time is created, because if you assume a 'past', it then becomes logical to think that there is a future, even though

114

the only evidence for future is the past, and the *past* is in turn based solely on the assumption that *memory-is-past*.

Derrida was explaining that if you say something of meaning, and if all of language is made of binary oppositions, then examining what was 'not being said' (looking at the opposite side of the duality) would give one a deeper understanding of the text. This is very interesting, and I think this is an important observation of how language operates. What he is saying is that all of our language and all of our linguistic meanings are constructs. Where Derrida went off the reasoned path, in his effort to understand time, is when he mistakenly forms a 'binary opposition' with 'past and future' on one side and 'present being' on the other. Perhaps we should turn our gaze away from linguistic theory and toward reality. Derrida got lost in the weeds, but he was astute in trying to diagnose what is wrong with time. Derrida got it wrong, but most never question as far as he did.

Chapter Fifteen

Nietzsche and Eternal Recurrence

Glance into the world just as though time were gone:
and everything crooked will become straight to you.
—Friedrich Nietzsche[1]

Nietzsche (1844–1900) does not present a systematic philosophy—in fact, his philosophy is often incoherent[2]—but at the center of all his philosophical thinking are his thoughts about time. Understanding how Nietzsche conceived time brings some order to the overall unsystematic array of his writings. Nietzsche's greatest concern with time was about the unalterable past. As Mircea Eliade and others have pointed out, all salvation is a salvation from time,[3] and the time that Nietzsche was seeking salvation from was specifically the *past*. The usual fear of time, from which most people want to be saved, is their inevitable death in *future time*. Some people obsess over the past more than they worry about their imminent death, but they are usually older people who know they have already experienced their best years, whereas Nietzsche was obsessing over his irredeemable past by the time he was in his late twenties.[4] Nietzsche wanted to redeem the past, and as a consequence, the construction of his temporal theory was tailored to address his dismay about the unalterable past.

Nietzsche's theory of time is referred to as *eternal recurrence* (or eternal return) and is based on several fundamental assumptions, the first being the assumption of time itself. He then assumes time extends infinitely into the past, and also that it extends infinitely into the future. He further assumes the material universe is finite in extent, and from this Nietzsche reasons there is only a finite number of states in which the

117

universe can exist, thus in the infinity of time, the finite number of possible states of the universe will be repeated endlessly. Historian Will Durant explains Nietzsche's eternal recurrence as follows:

> "The possible combinations of reality are limited, and time is endless; some day, inevitably, life and matter will fall into just such a form as they once had, and out of that fatal repetition all history must unwind its devious course again."[5]

The idea of a recurring cyclic time is not a concept that Nietzsche invented; it was an idea he took from ancient Greek thought.[6] But this kind of *time* was a nightmare for Nietzsche because he would be compelled to relive the past eternally, and for Nietzsche it is his revulsion of the past that motivates his philosophy. One further supposition allows Nietzsche to avoid the outcome of repeating the same life over and over, and this is the assumption that there is free will. Then, if *he wills it,* things can be different. This opens the way for Nietzsche to expound on the central tenet of his philosophy—the *will to power*.[7]

Writing on Nietzsche's sense of *being*, philosopher Joan Stambaugh states, "Nietzsche is able to say that existence is an uninterrupted having been."[8] Nietzsche has the standard cultural view of time that comes from thinking that the experience of memory is actually an experience of past. What we are (our self-identity) is composed of what we remember. We are our memory, and if we think our memory is representative of the past then our existence, for someone like Nietzsche, becomes the experience of "an uninterrupted having been." This is an understandable, yet strange perspective, because most people acknowledge their past as fixed, already determined history, without the great emotional disturbance that the past seemed to stir for Nietzsche. Most people live toward the future rather than allowing their lives to be oriented toward the past. To understand why Nietzsche has this odd perspective that focuses on the past it is important to cite a few biographical details about his life as a young adult.

Nietzsche's ability was recognized early, and upon leaving his formal education at the university he was appointed assistant professor of classical philology at the University of Basel, and the following year promoted to full professor at the age of twenty-five. He was definitely on the academic fast-track. It is important to note that in all of Nietzsche's writings it is impossible not to detect the oversized ego that stands behind the text.[9] Nietzsche undoubtedly had big expectations after his genius was recognized and then certified by his rapid ascent into the academic establishment. In retrospect, his academic career at this point had probably already peaked. He was plagued by health problems, and had difficulty in all his close relationships. Students avoided signing up for his classes, and he took several extended leaves from teaching for health reasons. He had a disdain, if not outright hatred of women, and (aside from a troublesome relationship with his mother and sister) he never maintained a close relationship with a woman. However, there is compelling circumstantial evidence that Nietzsche did contract syphilis while he was attending university.[10] When the symptoms appeared, he knew very well what the problem was, and at that time there was no cure—the end result was an inevitable descent into insanity. Nietzsche died insane, at the age of fifty-six. Was it this fate that focused his mind on the past, because for him the unpleasant future was known?

As Nietzsche saw it, his existence could only be redeemed if there was a do-over of the past, because the great promise that his young life held was otherwise lost. Nietzsche was a philologist who had carefully studied ancient Greek writing, and this is where he encountered, and subsequently adopted, the concept of eternal recurrence. Nietzsche has been criticized for appropriating eternal recurrence because it seems like a cheap grasping for salvation,[11] even though Nietzsche himself considered it one of his greatest insights.[12] Cyclic time opens up the possibility of repeating the past. If you assume only eternal return without free will the world is strictly deterministic, and even embracing your fate or, conversely, trying to alter it is already determined for all eternity. Eternal recurrence is a changeless, fixed universe and so it offers no real immortality.[13] Nietzsche was further impelled to take up his idea of 'will to power', for only then can there be the possibility of changing the past. To redeem the past it was necessary to have the power of intervening upon life's circumstances—free will was requisite for any real salvation that could be had in a world of eternal recurrence. But free will is also an unsupported supposition.[14] Nevertheless, these assumptions take Nietzsche on a speculative adventure that supposedly ensures him infinite do-overs, allowing him to exercise the changes that he wills.

Interestingly, Nietzsche acknowledges that there is a distinctly timeless aspect to reality and that a human can have an atemporal existence. He knows that the animal lives in timelessness:

"Man ... envies the animal, which immediately forgets and sees each moment really die, sink back into deep night extinguished forever. In this way the animal lives unhistorically ... "[15]

However, Nietzsche gets it wrong. The animal doesn't forget, it just doesn't get its memory entangled in the time concept. He does say, "Time in itself is nonsense: there is time only for a sensible creature."[16] And he also recognized that the young child abides in the atemporal, but thinks that the child is "blissfully blind" because he or she forgets:

"This is why he [the observer] is moved, as though he remembered a lost paradise, when he sees a grazing herd, or, in more intimate proximity, sees a child, which as yet has nothing past to deny, playing between the fences of past and future in blissful blindness."[17]

It is ironic that Nietzsche said that man is moved by seeing the child and the animal because it triggers memories of a lost paradise, but at the same time accuses the child and the animal of forgetting. The child's memory is functioning just fine. The reason the child is still in the garden is because her culture hasn't completed the temporal indoctrination. In fact, it is Nietzsche who has forgotten his own nature—he has forgotten his timeless memory, that sixth sense which existed prior to thinking that memory is a representation of the past.

Nietzsche did have some comprehension of a timeless being that wasn't attributable to the naivety of the simple mind of the child or animal. He saw that there was a way of being in which the adult mind could grasp the ahistorical, and he referred to this state of mind as the *suprahistorical*: "For the suprahistorical man the past and the present are one and the same."[18] He understood that the suprahistorical human stood against all forms of time. He said that the suprahistorical man stands "in opposition of all historical modes of regarding the past, they [the suprahistorical] are unanimous in the proposition: the past and the present are one ... ".[19] At some level Nietzsche knew that his understanding of time was obstructing his view of reality. Nietzsche writes: "One is vividly impressed with the very relative nature of all notions of time. It would even seem as if a whole diversity of things were really all of a piece, and that time is only a cloud which makes it hard for our eyes to perceive the oneness of them."[20] Nietzsche also knew that timeless-being was unattached to action and change. He said, "One could call such a standpoint suprahistorical, because one who has adopted it could no longer be tempted at all to continue to live and cooperate in making history."[21] However, Nietzsche had little interest in the suprahistorical perspective; he wanted to cooperate in making history and so he was focused only on the past.

Nietzsche quickly dismisses the timeless perspective:

"But let us leave the suprahistorical men to their nausea and their wisdom: today let us rejoice for once in our unwisdom and, as believers in deeds and progress and as honourers of the process, give ourselves a holiday. Our valuation of the historical may be only an occidental prejudice: but let us at least make progress within this prejudice and not stand still! Let us at least learn better how to employ history for the purpose of *life*! Then we will gladly acknowledge that the suprahistorical outlook possesses more wisdom than we do, provided that we can only be sure that we possess more life: for then our unwisdom will at any rate have more future than their wisdom will."[22]

Nietzsche is fully invested in time. He writes, "To imprint the character of being upon becoming—that is the *highest will to power.*"[23] But he also exhibits a passive acceptance: "He conceives of being as 'becoming' and as 'eternal recurrence' and reacts to it with '*amor fati*'."[24] Nietzsche's theory was that you had best learn to love your life because you have no other choice. But inherent in Nietzsche's concept of time, there is the *will to power.*[25]

Forgetting for the moment that Nietzsche's theory of time is a collection of cobbled-together ideas that create a tangle of problems, for Nietzsche it was sufficient ground to proclaim his maxim of will-to-power. Nietzsche, by invoking will-to-power, is declaring the will-to-act as the highest priority. He was making an argument that would later come to be associated with existentialism.[26] The existentialist proclaims that the most concrete fact is our own existence. They are right. Existence is the *given* that cannot be denied, and there is nothing else that can be known with such complete certainty. So the question becomes, what is to be done with this existence? How is it to be lived? Nietzsche was adamant that ego should take command of existence and exert ego's will on the world. The existentialist was trying to take control of his or her own existence, and this required an elan that embraced one's own fate or exercised the will to remake the world. Whether embracing or shaping one's existence, there is an undeniable nuance that assumes free will. Perhaps the Existentialist movement has all but died out, but its influence reinforces a very contemporary attitude which holds that, we humans can will-the-world, when in fact, by attempting to do so we are destroying the world.

A final thought on Nietzsche. Many defenders of Nietzsche have rejected the idea that his descent into madness was due to his illness. His defenders argue that he was so far ahead of his time that his madness was brought on by the society that surrounded him.[27] There is another possibility. Nietzsche's own philosophy did it to him. If will can make the world, as Nietzsche seems to think, and thus it is logically permissible to make assumptions, as Nietzsche did, then truth is cheap, and you can simply fabricate truth by believing in things. But this is a path of extreme danger. For instance, Nietzsche most admired the 'great man'—and he considered Napoleon to be a very great man. Doesn't it follow that one who is a great man, through his will to power, will become Napoleonic? The danger in thinking like this is that you might end up in a mental asylum fully convinced that you are a great man, in fact, convinced that you are Napoleon. Nietzsche did all this.

In conclusion, Nietzsche's brain was, in all probability, destroyed by syphilis, but prior to that his facility for reasoning was greatly compromised by that ever-present temporal virus, the sickness of time. He was infected with the culturally instilled virus, the belief that *memory-is-past*.

Chapter Sixteen

Longchenpa

When thy mind leaves behind its dark forest of delusion, thou shalt go beyond
the scriptures of times past and still to come.
—*Bhagavad Gita*, c. 500 BCE[1]

Longchenpa (1308–1364) is, "undisputably the greatest scholar of the
Nyingma tradition"[2] of Tibetan Buddhism. The Nyingma tradition is the
oldest lineage of Tibetan Buddhism, and was introduced in the 8th and
9th centuries from India. Longchenpa was a prolific writer, producing
270 works in his life. His major work is the *Seven Treasures*[3], which
brings a systematic and philosophical organization to the six hundred
years of Nyingma literature that preceded him, and culminates in the
exposition of Dzogchen. Considered the ultimate teaching of the
Nyingma tradition, Dzogchen is also known as Atiyoga.[4]

Longchenpa had a deep understanding of the time-self-memory-
nexus, and in the language of his own time and culture he successfully
unwraps the nexus. An explanation of Longchenpa's understanding of
Dzogchen and the time-self-memory-nexus requires a brief examination
of two technical terms; *trekcho* and *thogal*, which are the two main
aspects of Dzogchen practice. Trekcho is the practice that eliminates
time. Trekcho literally means 'cutting through', but as used in Dzogchen
it means "Cutting through the stream of the thoughts of the three times
[past, present, and future]."[5]

Thogal, the other technical term, is an expression of 'crossing over',
'leaping over', or 'passing over'—it is the change in mind from the
temporal to the 'naturally occurring timeless awareness'.[6] Whereas
trekcho is described as 'primordial purity', thogal is described as

'spontaneous presence'. Now, to put this in perspective, trekcho is identifying the root of time, understanding how that root is constructed, and severing the root, thus restoring 'primordial purity'. Thogal is the timeless mind that is 'spontaneously present'—when the root is cut, mind jumps out of time. Jumping out of time is 'cutting the root'.

Central to Longchenpa's philosophy is the denial of time. Of course, this is to be expected—it is axiomatic to every school of Buddhism that *time* is a constructed fiction,[7] but Longchenpa is exceptional in his close examination of the source of time. He demonstrates the requisite understanding that there is a nexus of intimately connected concepts, which form our sense of the temporal world. He knows the human *fall* into time corresponds to several interrelated ideas, including the *ego;* the *self-other duality*; the *schism of the senses*; and the *distortion of the sixth sense*. And there is the recognition that the distortion of the sixth sense is, in fact, a *distortion of memory*.

Longchenpa's denial of time is a theme that occurs throughout the text. He refers to reality (enlightenment) as 'natural occurring timeless awareness', which is the state of Being when mind is undistorted. Demonstrating his complete rejection of the reality of time he said, "Surely everyone knows that everything is timeless!"[8] But, what is most intriguing about Lonchenpa's philosophy is *not* his denial of time, rather, it is the way he describes the practice of eliminating time from our ordinary thinking. He is explicit in saying that cutting the root of time (trekcho) is removing normal recollection, because normal recollection is the sense that memory is being *recalled*. This creates the mistaken impression that memory represents the past. "Ordinary recollection and thinking vanish" when "one cuts the root."[9] The naturally occurring spontaneous presence (thogal) becomes the state of present being because, "there is no recollection; timeless awareness is free of the very basis of recollection."[10] Longchenpa does not mean that memory no longer functions. Instead, it functions in a way that the terms *recollection* and *recall* no longer apply. Memory no longer seems like it is the experience of *re-membering* or *re-collecting,* and since both terms allude to 'a time before the present', the terms are no longer applicable. He stated that eliminating the assumption that *memory-is-past* results in the situation where "Deliberate recollection is pacified and there is no conceptual elaboration."[11] In other words, memory is always nothing other than the present experience of memory. "One who brings an end to recollections of the past, curbs anticipations of the future, and allows consideration of the present to fade naturally is what I call 'a yogin who knows the equalness of the three times'."[12] Having cut through the root of time, timeless awareness naturally occurs. "Since nothing that involves deliberate recollection applies to me, I am liberated from the painful phenomena of samsara."[13]

From the above it becomes clear that Longchenpa knows that memory is infected with a misunderstanding, and he is aware that the root of time and the infected memory are the same problem. Longchenpa sees the consequences of time as the human problem, and sees the solution as 'cutting the root' of temporality. He further recognizes that 'cutting the root' obviously entails removing the supposition of 'recollection'. There is memory, but there is no 're-membering', because that requires that memory is assumed to represent past. Past is a fiction, and if past is a fiction, so too is time a fiction.

One of the most compelling aspects of Longchenpa's philosophy is the *practice*, which if practiced correctly, leaves nothing to do. In complete opposition to Nietzsche's will-to-power, Longchenpa advocates no action. Longchenpa explains, "As for discerning timeless awareness, it is realized to be awareness free of plans or actions,"[14] and "It is never necessary to engage in enlightened activities that involve effort."[15] He also stated that "nothing need be done,"[16] because "Deliberate action misleads" and "Effort corrupts."[17] Keith Dowman elaborates on the practice: "Dzogchen stresses the undeniable fact that any goal-oriented conscientious endeavor assumes a result in a future that by definition never comes and thereby precludes any attainment in the present moment.

Thus there can be no liberation until the drive to attainment is relinquished ... 'Nonaction' is the salient key term in the evocation of reality of natural perfection."[18] This is a problem for the ego, but correct insight will lead to the dissipation of the ego. The practice, once you have achieved insight into the nature of the deception, becomes a present doing without any further ambition.

Longchenpa's rejection of sitting meditation[19] carries non-action to its logical conclusion—concluding that sitting meditation is also a forced action. No doubt, sitting can quiet the mind, but after you have reached a point of quiescence, you must use this quietude to focus on *how* mind became entrapped in its own confusion. Here the practice of Dzogchen parallels the practice of philosophy, and Longchenpa was, without a doubt, a philosopher of the first rank. Longchenpa makes it clear that it is our own mind that is our teacher:

> "When I, the majestic creativity of the universe,
> Manifest as the teacher, your own mind,
> You should listen to this message: your own mind is the teacher."[20]

He projects a clarity of thought that stands out in the Buddhist literature—he is as direct in his treatment of phenomena as was Nagarjuna,[21] but his understanding of time surpasses even what Nagarjuna knew. Longchenpa wrote with a directness that makes it hard to ignore or misinterpret what is being said. This does not mean that what he said is easy to understand. However, there is no mistaking that he considers time a fiction, along with the ego, and recollection. His emphasis on recollection is not a coincidence; he knows that memory is entangled in the illusion of temporality. He understands that ego is a construct, dualistic thinking is the source of illusion, and that the illusion is due to a distortion of the six senses—a schism between mind and body. And further, he knows that the distortion is of the sixth sense; in particular, it is a distortion of mind, specifically memory.

A single sample of the compact and intense writing of Longchenpa will serve to illustrate his direct and forceful style. He can deny *self/other* and the *mind/body duality*; *time and space*; and *beginning and end*, all within one short sentence:

> "Self-knowing awareness, involving no perception of outer objects
> and inner subject,
> has no time and place and is beyond phenomena that originate
> or cease."[22]

Summary: The many thinkers and philosophers who have considered time offer a myriad of arguments for, and explanations of, time phenomena. Everything that could be said about time seems to have been said,[23] except the thesis of *Time Sutra* which has been overlooked—the thesis that declares our experience of the world is badly distorted due to our universal assumption that *memory is past*. There seems to be little understanding that so many of our problems flow from this single mental error about time. But all of the elements necessary to understanding time are present somewhere in the ongoing and wide-ranging debate about time. And the understanding is actually quite simple: there are only three possible experiences of time: past, present, and future. The past has been shown to be unreal. And no one has ever experienced the future. Thus the present is our complete experience of the world. Without either a past before now, or a future after now, the present is *timeless*.

Modernity, by its very name, implies time. But even if reality is timeless, there is something to be learned from modernity. Before we turn the page and take up the postmodern, we need to look more closely, otherwise we have learned nothing over the past four centuries. There is a lot of truth in the position of the radical postmodernist who thinks we have ventured four hundred years down a blind alley. But this is only a diagnosis of the problem; the postmodernist doesn't know the cause, and has no cure. The whole human debacle is due to the formalization of time, yet, this same formalization, carefully studied, brings into clear focus how humans wrongly think that their memory is an experience of a time that belongs to the past.

Part V

The Path

Chapter Seventeen

Abyss

It is not doubt, it is certainty that brings on madness ...
We are afraid of the truth ...
—Nietzsche [1]

The modern digital computer is often argued to be analogous in many ways to the human brain. This is a useful analogy, because to the extent the comparison holds, it might provide some insights into consciousness. However, consciousness is exactly the point where the computer analogy meets with the most resistance. Most would agree with the idea that the brain functions in some respects like a computer. The brain functions as a memory with the capacity to record and to recall. The brain, through conditioning/learning, undergoes a process that is similar to the programming of a computer—information is encoded into the brain, and revealed in output in the form of behavior. The most controversial issue is whether consciousness itself can arise in a sufficiently sophisticated computer. Those scientists and philosophers who cannot accept this possibility tend to reject the idea that consciousness is reducible to the laws of physics, chemistry and biology. They dispute the purely materialist viewpoint, and insist that mind possesses some additional, but unspecified property. There are two arguments for this view, and they are not necessarily separate. Some hold that out of complexity, an irreducible property emerges that is inexplicable at lower levels of organization. This is the argument for *emergence*. Others argue that mind is truly a substance different from matter, thus the world is split by the mind/matter duality. This view is often referred to as *substance duality*.

This debate, pitting the materialist/reductionist against the advocates of 'emergence' or 'mind substance,' is very similar to the argument put forth by *vitalists* in the 19th century and first half of the 20th century. Vitalism maintained that there was a 'life force' in living organisms that was not reducible to physico-chemical processes. Here it was argued that life is an inexplicable emergent property that appears out of complexity, or there exists some mysterious life-animating substance that is present. The argument, for the most part, was over by the mid-twentieth century. In the 1950s vitalism was empirically refuted by the discovery of the DNA structure and the subsequent deciphering of the genetic code.

The emergence argument was half right—life arises out of complexity, but it was wrong to argue that life couldn't be understood at the physical-chemical level. There was no secret 'life force' other than a detailed set of chemical instructions. Each genetic code (genome) is essentially the 'instruction set' for constructing a particular member of a species of organism. It would be difficult to find many in the biological sciences who still believe that some additional 'life force' is necessary to animate an organism. Presently genetic engineers create new organisms, though not yet from scratch, but through manipulation of the genetic code. Life-forms are described by the information contained in their respective genetic sequence. We humans are, in a sense, not only the sum of the total information contained in our genetic sequence, but also the total of the information encoded in our neurons—information both hardwired and learned.[2] The vitalists lost their argument and their present day counterparts, the emergentists and the dualists, will probably lose their argument—an argument that tries to make consciousness into an unknowable property.

The reason the brain looks a lot like a computer is that the human brain, whether knowingly or unknowingly, intended to design a computer to function in ways that the brain does. The fact is, the brain made the computer so that it would perform like the brain. The computer is just another tool made by humans to allow a task to be performed easier, or faster, or more reliably. Many of the tools that humans construct are designed to extend our senses or give our limbs, hands, and feet greater ability. The computer is a tool that extends the ability of our sixth-sense. It enhances the capability of memory in two important ways, giving us potentially unlimited, reliable storage, and supplying the facility to make incredibly fast and accurate calculations.

Norbert Wiener (1894-1964), the originator of the word cybernetics,[3] in one of his writings on cybernetics said, "It is an interesting reflection that every tool has a genealogy, and that it is descended from the tool by which it has itself been constructed."[4] In the sixty years since Wiener made this observation the digital electronic computer has developed an extensive genealogy—significant technical advancements sometimes lead to what is called a new generation of computers. Computers are used to design newer computers. The complex designs of integrated circuits are computer designed and the manufacture of those integrated circuit boards and chips can only be produced through precision control by computers. The computer is a tool used to make an ever more refined tool. Perhaps the greatest tool that serves as an extension of memory is the Internet, which is to date the ultimate computer tool.

We have a rough understanding of how the brain, as a piece of hardware, operates. However, the real value of the computer analogy is in understanding the software that makes the brain function as it does. If we are to pin down the elusive self-consciousness it will not be in the hardware, it will be contained in the software. Daniel Dennett made the brilliant leap in understanding that the self-conscious ego is not a product of biology per se, but is constituted out of software. Just as we can read and comprehend the DNA sequence as the source code for a living organism, and thus do not have to invoke *vitalism* to explain life, so it is the same with consciousness: if you can understand the software inscribed into memory, you can understand the appearance of the conscious self-ego.

The computer analogy leads to the understanding that the way we think is based on what is encoded in the mind. If the encoded instructions don't agree with reality then it introduces a certain craziness into the way we think, and even a casual observer of human behavior can see that there is a lot of craziness out there. Few would disagree, but it is always someone else's madness—people are blind to their own nuttiness.

The mind, the sense of all that is mental, is attributable to the functioning of the brain, and the brain's function is similar to a computer. The brain has an operating system that is genetically hard-wired in the parts of the brain that are of older evolutionary origin. But in the newer parts of the brain, where most learning occurs, our childhood cultural indoctrination lies buried. All of us have a cache of unacknowledged, and thus unquestioned, beliefs that filter our view of reality. If you can exhume the most deeply buried beliefs, those that most distort reality, there will be a dramatic shift in the way you perceive the world. The very root of ego and the root of our delusion is derived from a belief into which we were indoctrinated. From the time we were infants we were taught to think that *memory-is-past.* This one supposition makes us think the world is temporal, and that we are a being bracketed by birth and death. But if the computer analogy is correct in the way that it is herein presented, then it indicates that not only memory—the brain's hard drive—has been corrupted, but the brain's operating system has been hijacked by a virus that imposes the temporal view.

If we want an explanation in cause and effect terms, the computer analogy works well. It tells how it is that we might acquire an egoic, temporal-self. It explains that we assemble self-ego out of memory. If you are looking for truth, then the code that is culturally implanted in memory, a code that unconsciously assembles the self, begins to look exactly like a computer virus. This "virus" is carried in our culture, so the virus infects the brain of every member of our culture from an early age. To accept this verdict is to doubt the reality of the ego, and to doubt the ego is to approach the abyss.

It is the realization of the insubstantiality of the self that affords the first look at the abyss. For most, that first glimpse is threatening enough to cause them to turn away and never look back. Huang Po, the 9th century Chinese Zen master, describes looking into the abyss: "It is neither subjective nor objective, has no specific location, is formless, and cannot vanish. Those who hasten towards it dare not enter, fearing to hurtle down through the void with nothing to cling to or to stay their fall. So they look to the brink and retreat."[5] The appearance of the abyss is all about doubt, and the doubt of the self is the most disturbing possible doubt. This doubt invokes a feeling of depression and a fear of nothingness.

The great American philosopher William James writes of the distress that is felt on looking into the void: "The apprehension was a coming dissolution, the grim conviction that this state was the last state of the conscious Self, the sense that I had followed the last thread of being to the verge of the abyss, and had arrived at demonstration of eternal Maya[6] or illusion, stirred or seemed to stir me up again."[7] This is the fear of the complete disintegration of the self. This scares most everyone away, but they are mistaken in their fear. They have mistaken the temporal self-construction to be reality rather than seeing the original primitive being that lies at the beginning, before time. It is the path of doubt that has access to that timeless realm, but if you are to travel this path, the ego must confront the abyss and a successful confrontation will always be fatal to the ego. The Zen tradition often refers euphemistically to this confrontation as Great Death[8]—this is the death that liberates being in such a way that the once feared death-in-time is of no further concern.

The Western Cartesian approach to the path of doubt was established nearly four centuries ago as a formal way of doubting, but after all this time we have not doubted all the way through to the very end of modernity. Many have wanted to get beyond the mess that modernity has foisted on the world, and so they propose the *postmodern*[9], but they have never learned the lesson of modernity. Learn the lesson of modernity and there will be no need of the postmodern. Postmodern, by its very name means *'after modernity'*. Those who are proposing the idea of the *postmodern* do so with the intent of ending modernity by relegating modernity to the past—they do this because they never learned that the 'conceptualization of *past*' is the error that created modernity. Ending modernity by trying to cast modernity as the past, as the postmodernist does, is just another way of inadvertently accepting a mistaken view about the existence of time. *Time Sutra*, like postmodernism, also has the intent of ending modernity, but this path to the end of modernity coincides with the end of time—memory is *not* past, and there is nothing before or after. *Modernity* and *time* end, because neither are real.

Time Sutra is the practice of doubt, and to successfully practice doubt, all the barriers that belief erects must be overcome with doubt—particularly the barrier that is erected by the belief in the self-ego. This path demands sense, reason, experience and logic, but it also insists on the courage to face Great Death—the loss of ego. This requires the courage to look into the mirror of consciousness and see that there is no ego reflected. As the Buddhists would say, the mirror is empty, it reflects everything, but there is *no self-reflection*. Buddhists, for twenty-five hundred years have made the elimination of the ego the central intent of their practice. Alexandra David-Neel, a Buddhist practitioner and the first Western woman to travel to Lhasa, Tibet, describes the "Direct Path" as the steep path "that is dangerous because the guard-rails of social, moral and religious laws are missing from its edge, and the traveler who makes a false step risks a fall into the abyss."[10] Ancient Zen master Huang Po understood the risks and offered the following advice:

"Bravely let go on the edge of the cliff.
Throw yourself into the abyss with decision and courage.
You only revive after death."[11]

Chapter Eighteen

Modernity's End and The Path of Doubt

*The search for reality is the most dangerous of all undertakings
for it will destroy the world in which you live.*

—Nisargadatta [1]

From its foundation in Cartesian rationalism and Baconian empiricism, the modern spacetime continuum is a philosophical construct that has evolved for nearly four hundred years. The demand that the system conform to empirical observation necessitated that the system change by adapting to new empirical discoveries. Occasionally new theories or scientific findings would create an upheaval that resulted in a 'paradigm shift.' In the words of Thomas Kuhn[2] the paradigm shift was a "scientific revolution" which alters the entire system when the older way of thinking can no longer be reconciled with the new data/observations. Kuhn made the point that the growth of scientific understanding wasn't a smooth and gradual evolution—the new paradigm was often incommensurable with the old paradigm. However, during the time in which a paradigm is in effect (often referred to as the 'ruling paradigm'), its supporters always consider it both rational and empirical. Morris Berman, speaking of our present day 'ruling paradigm' writes, "The reigning worldview ... [is] modern science. The scientific establishment of the twentieth century is the real Church of the West."[3]

The argument that *Time Sutra* puts forth is not a call for a paradigm shift, it is more radical than that. To paraphrase Namkhai Norbu, we don't want to build another, better cage, rather, we want to replace the old cage with no cage at all.[4] The empirical nature of the modern scientific paradigm demands *time*, but time is nowhere to be found in the physical world, except as an assumption about memory (in physics it has been named The Past Hypothesis, and in philosophy it has been called The Representative Theory of Memory). Physicist and philosopher David Bohm said that "Progress in science is usually made by dropping assumptions."[5] But in this instance the assumption that must be dropped is time, which is fatal to the whole enterprise of modernity. In the complete absence of time, the modern scientific paradigm cannot exist. The rational-empirical foundation on which modern science is grounded logically leads to its own demise. The age of modernity grew out of the logic of the 17th century scientific paradigm, and that logic has now been followed to a logical end. Modernity is eliminated by the same Cartesian method-of-doubt that foisted modernity on the world. Just keep the doubt going until you reach the end of belief and you simultaneously reach the end of time. What could be a more convincing end-to-modernity than the ending of time, the very substance out of which modernity was created.[6]

We should not expend effort looking for a better paradigm—it doesn't exist. The philosopher and historian R. G. Collingwood explains that there can never be a constructed system, like the scientific paradigm, that embodies truth. He writes, "The truth is not some perfect system of philosophy: it is simply the way in which all systems, however perfect, collapse into nothingness on the discovery that they are only systems."[7]

This truth did not take four hundred years to be realized—it can be realized at any time and place. In fact, over fifty years before René Descartes published *Discourse on Method* in1637, and thirty-nine years prior to the publication of Francis Bacon's *Novum Organum* in 1620, Francisco Sanches (1550-1623), in 1581, published his treatise *That Nothing is Known*.[8] It is almost certain that Sanches influenced Descartes' thinking.[9] Sanches opens his treatise with the declaration, "Let this proposition be my battle colour—it commands my allegiance—'Nothing is known'."[10] And he later states: "Let us come to our conclusion. All understanding is derived from the senses, and beyond this kind of understanding, all is confusion, doubt, perplexity, guesswork;

nothing is certain."[11] Descartes' thinking, by comparison, is timid and less radical than Sanches' philosophy. But Sanches was right and Descartes was wrong. Descartes (perhaps knowingly) was afraid that he would lose his God and his self-ego if he allowed unrestricted doubt. The fear of 'not knowing' intimidates the ego and restrains our practice of doubt—the ego is composed of what it knows and what it knows is contingent upon beliefs. The ego's very existence is dependent upon its believed-in knowledge. Doubts can only persist in the presence of belief—belief comes first and then doubts arise in response to the supposition of the belief. When there is no belief present, doubt is logically impossible. At any time and place this *reason* is available to everyone.

Albert Camus said it best, "Our reason has swept everything away. We build our empire upon a desert."[12] This is the despair of the self-ego when it loses all claim to knowing anything at all. The argument that demonstrates the truth of Sanches' *That Nothing is Known* will always be unacceptable to self-ego. So here we are—stranded on the path—stunned by confusion. This is where the path steepens—just knowing that the self-ego is a fraud doesn't make it go away. There is, however, a domain of experience and wisdom that has already cleared a pathway. Buddhism has spent twenty-five hundred years confronting the self-ego and trying to make ego disappear—this is valuable experience that should not be overlooked.

Sixteen centuries ago in China, Bodhidharma, the originator of the Zen tradition, said the following: "There are many roads that lead to the path, but basically there are only two: reason and practice."[13] The path of *Time Sutra* is the path of *reason*, so it is only natural that Buddhism should be consulted about the path which Buddhists have followed for the past two and a half millennia. In Buddhism it is axiomatic that the ego be overcome—removal of the self-ego is the foremost requirement on the path to enlightenment. Before taking up the path of *reason*, it is important to say something about the alternative path that Bodhidharma mentions—the path of *practice*.

Path of Practice: Practice in Buddhism takes many forms, and as the term *practice* implies, there is a discipline of habit, which is intended to change the way we behave, and by extension, change the way we think. Some of the habits that Buddhist practice promotes are acting with tolerance, love and patience—in other words 'acting Christlike'. This is a training to act selflessly, which might help but doesn't remove the ego. This is certainly a socially positive practice, and it is claimed by its practitioners to open up many insights. Another type of practice that is nearly universal in Buddhism, is stoic sitting-meditation—an attempt to bring some quiescence to the mental noise that fills the stream of consciousness. This sitting-still 'path of practice' fulfills the need to quiet the internal chatter, and does so by closely observing our own mental activity. By recognizing the endless nonsense that commonly occupies the mind at every moment, and then focusing continuously on this concern, through practice, the mind clears enough to glimpse enlightened insight.

Most practitioners who have success at sitting meditation confront another problem when they stand up from the meditation cushion—they quickly fall back into the temporal world of illusion. In 2nd or 3rd century India, Nagarjuna reflected on this problem, " ... yet when a being is no longer meditating, suffering will be experienced because all the ignorant views and karmic formations will regain their power over that being. This is because they are merely suppressed during meditation ..."[14] Without a deeper understanding of the cause of our fall into unreality, sitting meditation can only bring temporary relief from the pain of our mixed-up being. Many of the great Buddhist masters advocated sitting-meditation to still the chaos of thought, not as a method to eradicate thought, but rather, while the mind is quiet, to examine the nature of mind. From this point of understanding, mind switches from practicing a habit of dispassionately monitoring the stream of our thoughts, to cultivating an intense attention to the reason that drives our thoughts. This is also the point where the path, having already established the habit of practice, enters onto the path of reason.

A final comment on the 'path of practice' is necessary. There is a common trap associated with the practice of sitting meditation. The attainment of complete quiescence can turn into oblivion, resulting in a fall into complete nothingness, a sort of forced catatonic-like state. Buddhism has long been unjustly accused of promoting nihilism[15] for wanting to kill the ego (of course the criticism comes from ego, it is always the self-ego that makes the loudest objection).[16] However, when it comes to this particular error of sitting meditation which promotes oblivion, the consequences really can be complete nihilism, in that *being* itself is lost in non-being. This type of practice is most often seen in Zen practice.[17] Sitting-Zen (zazen) sometimes takes on the form of complete quietude without even thought—this happens as a result of the mistaken attitude that it is not a problem with the 'way we think', rather, *'thinking itself'* is the problem.

The reality is that humans think—they think a lot. This is what humans do. Culture is made of language, and it is thinking in language that makes culture, and our culture makes us what we are. Human thinking is generally confused, but stopping thought altogether is not the answer. Humans are the animals that *think*—this is our nature. Learning to 'think clearly' is the answer, and achieving clear thinking will likely require a discipline of practice, but in the end practice must be accompanied by *reason*.

Path of Reason: When memory/mind becomes quiet, then it becomes possible to quietly observe the sixth-sense, which is no longer completely owned by ego. American poet, writer and practicing Buddhist from the Beat Generation, Gary Snyder describes the experience: "Meditation is going into the mind to see this [intuitive wisdom] for yourself—over and over again, until it becomes the mind you live in."[18] But to do this requires a solitude that sometimes creates a distance from others. The duration of quietude necessary to follow this meditation tends to isolate anyone who follows this path.[19] However, the efficacy of the path is without question, since this is essentially the same route of all the buddhas.[20]

The path of reason is easy to understand, but it is a path that is hard to follow. All that the path requires is that you lay down all your assumptions, and that which remains, is reality. However, humans seem far more inclined to believe than to doubt, and for this reason bringing sufficient doubt to bear on our most treasured beliefs is, for most believers, not possible. But acquiring the courage to doubt is not the only obstacle on the path. Equally daunting is the isolation and solitude necessary to traverse the narrow, steep path of reason.

Frequently, in the Buddhist literature there is an insistence that the follower of the Buddhist path must find a teacher and establish a guru-disciple relationship. Counter to this, there is also a strong tradition in Buddhism that began with the Buddha, which insists the path be taken alone. Competent teachers are rare, but an alternative—the written words of the ancient masters—is available to anyone who looks. When Muso Soseki (1275–1351) was a young Zen practitioner he was given this advice by an elder, " ... it is very hard to find true teachers. But if we read the records in order to encourage our own earnest aspiration, we will come to see the forerunners' satori[21] experiences are really our own, here and now. Where in the world, then, is the difference between past and present?"[22] Hui-neng (638–713), the 6th Patriarch of Zen, said, "If you insist that a teacher is necessary to attain liberation, you are wrong. Why? Because there is a teacher within your own mind who enlightens you spontaneously."[23]

Insight attained on the path of reason is not something that can be given to you by someone else. Noted mythologist Joseph Campbell stated that, "Since in the world of time every man lives but one life, it is in himself that he must search for the secret of the Garden."[24] Keizan (1268–1325), a second generation heir to the teachings of Zen master Dogen (1200–1254), said "Zen study demands that you investigate and awaken on your own."[25] Chogyam Trungpa (1940–1987), a contemporary Tibetan master who taught in the West for many years said, "We have to accept that ours is a lonely journey ... the whole journey is made alone, independent of anybody else."[26]

The most pleasurable aspect of meditation on the path of reason is following a train of rational thought that ultimately alludes to that which cannot be thought. Humans are thinking animals and one of our most pleasurable activities is thinking—we love our thinking, but if we *think* based on unexamined beliefs, we are risking our own sanity. There is, however, a philosophy that adheres to logic, and there is really only one logical philosophy possible, but the narrative of this philosophy is told in a myriad of ways. It is sometimes referred to as the Perennial Philosophy[27]—of which Buddhism is one species and *Time Sutra* is another species—same philosophy, different narratives.[28]

Chapter Nineteen

Meditation on Doubt

From the beginning there's not a single thing before or after.
—Dogen, 13th century CE [1]

The path of doubt is characterized by a critical attitude which is known in philosophy as skepticism. Skepticism is an attitude of questioning beliefs and assumptions, an activity that is really at the core of what philosophy does—philosophy is always a form of inquiry, so it is hard to think of philosophy as separate from skepticism. However, radical philosophical skepticism (i.e. complete or total skepticism) is generally dismissed by most philosophers because total skepticism is an argument easily deployed against every form of speculative philosophy. And most philosophers indulge in some degree of speculative philosophy. They allow their skepticism to penetrate only so far, stopping before their skepticism confronts their deepest beliefs. The great majority of philosophers would see this 'meditation on doubt' as complete nihilism—but it only seems like it, because it is the annihilation of the worldview of the ego.

Philosophical skepticism is a very old tradition that permeates philosophy in both the East and the West. Skepticism can be traced in the West back to the ancient Greeks. Pyrrho of Elis (c. 365–c. 270 BCE) is regarded as the founder of skepticism,[2] but Pyrrho was preceded by the skeptical philosophy of Parmenides of Elea (born c. 515 BCE). Parmenides was the most radical skeptic of the pre-Socratic philosophers, but others at about the same time were displaying the skeptical attitude, including Democritus (460–370 BCE), Xenophanes (570–475 BCE), and Heraclitus (c. 500 BCE).

In ancient India, the skeptical tradition is at least as old as Buddhism,[3] but even Buddhism is preceded by the older Vedas, some of which adamantly doubt our conventional perception of reality—thus displaying a deeply radical form of skepticism. One of the oldest recorded expressions of total skepticism comes from Lao Tzu, the founder of Taoism, who lived in 6th century BCE China. Lao Tzu stated "To know that you do not know is best."[4] The long tradition of skepticism has been populated by many great philosophers who have all practiced a method of doubt. Those who have chosen to follow the skeptical path see skepticism as the only line of thinking that consistently employs a rigorous reason.[5] The extent to which you deviate from this path is the extent to which you abandon reason.

Traditional skepticism is well established—it penetrates into every area of philosophical thought, from the very beginning of philosophy up to the present. There has never been a philosophical assertion of knowledge that hasn't been rebutted by someone who doubted that same assertion, and that doubter is to a degree a skeptic. In fact, a radical skeptic will argue that skepticism *is* the philosophical tradition and all else is dogmatism of one form or another. The present day embodiment of skepticism is best exemplified by postmodern thought. Postmodernism presents a forceful skepticism, but the postmodernist never understood the temporal ground of the modern era.

Modernity began, if we date it to the inception of Cartesian methodology, as a rigorous skeptical discipline that was intended to find the unshakable ground of reality. Although the scientific philosophy that evolved out of the Cartesian system was increasingly refined over nearly four centuries, it has always stopped short of doubting all the way down to the most fundamental belief. When philosophers of the 20th century began critically reevaluating the modern paradigm, which had now garnered worldwide acceptance (no other philosophical system had ever before dominated globally), it became obvious that no ground had ever been established. All that could be offered in support of modernity were unverifiable suppositions.

How did modernity, which was founded on a method of skepticism, for so long hide from the fact that its foundation is groundless? Descartes himself had numerous unacknowledged assumptions that were apparently beyond his capacity to doubt. In the intervening years, from Descartes' time up to the present, philosophers and scientists uncovered numerous suppositions, but there was an invisible boundary limiting doubt. The barrier to attaining complete doubt is a symptom of the human condition; we cling to our beliefs because we think that the self is contingent on the beliefs it holds.

Early in the modern era, for over two hundred years after Descartes, those who were fabricating the modern paradigm thought of themselves as Natural Philosophers, but at about the turn of the 20th century those who were formulating the modern paradigm began to refer to themselves as scientists.[6] Scientists no longer saw themselves as working within a philosophical frame of thought, and relegated philosophy to metaphysics, which was deemed a form of mental speculation that was often shunned by scientists. This shift in thinking altered the worldview of the ruling paradigm. Instead of a natural philosopher studying nature, the modernist became a scientist learning how to engineer nature. There have always been engineers of nature (alchemists trying to turn base metal into gold), and there have always been scientists/philosophers who are willing to doubt their own ground of self and time.[7] But the current balance is tilted toward belief rather than doubt, and it is belief that has proven itself dangerous.

The ruling scientific paradigm established itself as the unquestioned authority of truth, but the intervening events of the 20th century precipitated the re-emergence of doubt. The disaster of the First World War, followed two decades later by World War II, made it obvious to some that modernity was creating a dystopia. By the end of the 20th century the accumulating environmental disruptions were seen as clear warning that modern technology, long-term, poses a threat to most species of living things.

The postmodern thinker is right to question the path we are on, but postmodern thought is ill-equipped to follow through on a skeptical path. Of those closely associated with postmodernism, Jacques Derrida (1930–2004) stands out, and his philosophy demonstrates both the success and the limitations of a postmodern critique of modernity. Derrida and many others in the postmodern movement are of the continental philosophical tradition; they have never accepted the truth claims of empirical modern science. Derrida relies heavily on linguistic analysis, which provides him with an insight into how our interpretation of time is key to our worldview. Although Derrida recognized that time is key (he then stated that there is no evidence for time)[8] he nevertheless proceeds to reify time. David Loy points out that Derrida's time is just another constructed time.[9]

Because the great majority of postmodern thinking came out of the non-empirical mind-set, it prevented a deeper analysis of what four hundred years of ever closer observation reveals about time. Yes, everything is self-referential, and we are forever within the constraint of the hermeneutic. But, even though science is constrained to measuring phenomena against itself, it has a multi-century record of increasingly finer measurements of just about everything that could be measured. The most intriguing aspect of this treasure of experience is that everything is somehow related to everything else and with remarkable mathematical preciseness. The mathematical preciseness only breaks down when measurement is no longer precise.[10]

We effectively have, in the case of modernity, a four hundred-year experiment, but to a great extent the results haven't been as closely examined as the wealth of data would require. However, the postmodern philosopher is not properly prepared for, or even interested in, performing the required examination. Furthermore, beyond this 'four hundred-year experiment', humans have logged 2500 years of skeptical philosophizing in an effort to escape belief. Analysis of both the intricate observations of modernity, along with an examination of the expansive storied tradition of skeptical thought, is an opportunity to acquire insight into where things went wrong.

Postmodern skepticism is correct in seeing modernity as the road to disaster, but it is actually the observations and the theories of modern science that make the strongest case that we are on the wrong path, so the postmodernist has really added nothing new. Postmodernism offers no clearly articulated alternative to modernity, rather it tends to blunder forward, believing in the existence of a temporal world, not realizing that temporality is the problem. What the postmodern skeptic misses is that a deconstruction of modernity, when carefully considered, will expose how we have come to deceive ourselves. The postmodernist tries for a solution, but doesn't know how to fully carry out a deconstruction of modernity. The postmodernist accepts the time-space framework, but then rejects the empirical consequences. Jacques Derrida, who mastered the theory of deconstruction, tries to deconstruct modernity, but he becomes totally confused about the nature of time (saying we favor the present at the expense of past and future—thus he allows the fiction of time to stand, which happens to be the central feature of modernity, and the feature most in need of deconstruction).[11] The only possible correct deconstruction goes directly to the dismantlement of the time-space paradigm. If you fail to correctly deconstruct the central organizing principle of modernity, that is, *time and space*, you will fail to understand modernity.

Skepticism, followed to its conclusion, delivers an insight that is a clear vision of the world, uncontaminated by beliefs. The insight brings with it a calmness that is derived from an acceptance of being as it is. A quietude descends over mind because all the machinations that fill consciousness come to an end. Most of the churning of thoughts is a direct consequence of temporal thinking. It is common knowledge that thoughts about past and future fill many of our conscious moments. We think a lot about the future and it stirs emotions of fear and greed that wouldn't otherwise be aroused. Thoughts of past bring to mind unmet expectations and stir uncomfortable emotions of regret and anger. Insight into time absolves us of all of this fear, greed, regret, and anger, because time is no longer a constant consideration.

We are far greedier when we have a strong concept of future, and our anger and regret about our personal situation is greater when past expectations don't match reality. Temporal thinking agitates the mind into considering a plethora of possible future outcomes, and promotes a constant mulling-over the many lost opportunities. Buddhism refers to this problem as the 'three evils', often defined as 'greed, anger, and ignorance'. It is easy to understand how the unknown future can be something that is of great concern, and also how a certain grasping type of greed comes from wanting to make things better for the future self.

Skeptical traditions have commonly promised a tranquility, or an *ataraxia*[12] that comes with complete doubt. The Enlightenment of Buddhism is the unperturbed tranquility of attaining Buddha Nature. When you realize your own Buddha Nature, you become who you really are. The earliest writings of Taoist philosophy already exhibit a philosophy of tranquility—when the Taoists first encountered the Buddhists, they already knew what the Buddhists were saying. Some of the early Greeks saw skepticism as the path out of anxiety and unease.[13] When you no longer hold tightly or defend passionately your most cherished ideas, you begin to enter a state of tranquility. In the third-century, Sextus Empiricus, who practiced a radical skepticism, argued that dispensing with dualistic thinking will engender the tranquility that we have been seeking.[14] By knowing reality there appears a natural tranquility that is our original self. All of our ill-at-ease and dis-ease is caused by our existence in the temporal realm of samsara.

It is not a coincidence that both Buddhism and skepticism claim to achieve a state of tranquility, because both states of tranquility are arrived at in the same way—through the practice of doubt. Both Pyrrho and Sextus Empiricus made the case that the path of doubt will attain a state of tranquility that is reached when dogmatism is neutralized through skepticism. Skepticism offers the peace that comes with letting go of the self-ego's knowledge—the self is transformed into being-present. Buddhist teacher-practitioner Joseph Goldstein said it well, "This not-knowing is not a quality of bewilderment, its not a quality of confusion. It actually is like a breath of fresh air, an openness of mind."[15]

The entire philosophy of skepticism is quite simple, but there is a lot of explanation needed to justify embarking on such a seemingly radical path. The human condition is such that we have believed ourselves into an illusion that we don't know how to escape. The vast majority never even suspect that their worldview is an illusion. And the only way out is through the ego, which will not allow us to pass through, because we mistake ego for the true self. The argument made by the ego is that without ego there is nothing—and this is why it feels so nihilistic, but this ignores the fact that ego is a fabrication associated with the creation of time, which is in turn fabricated out of memory. The argument is persuasive that the self-ego is *the* construction that blinds us to reality. Only when we look, can we know. The path of doubt appears to risk oblivion, but in fact, the risk is already here—oblivion in time is certain—the consequence of time is death.

By deploying skepticism against the dogmatism that infects most humans, a tranquility comes over one who has attained the skeptical view. Skepticism and dogmatism are obviously just another manifestation of the doubt/belief duality. Dogmatism is a consequence of the 'set of beliefs' that one chooses to hold, and skepticism is the philosophical method of doubting that same 'set of beliefs'.

The self-ego, contingent on the knowledge contained in memory, is found to actually know nothing, and so the entire set of criteria for the self-ego has disappeared. If you can follow the gist of this argument, then the logical soundness is apparent. Although it is transparently obvious that belief is a logical fallacy, few understand the disastrous consequences that accrue from our most fundamental beliefs. Our true being is diminished by belief.

The last, and the most difficult task of skepticism is to doubt time, which requires exposing the foundational, temporal-assumption that *memory-is-past*. Modernity lays out the arguments for time in exquisite detail, so if there is to be a rejection of the temporal, the modern paradigm is where to look.

Part VI

Consequences of Time

Chapter Twenty

Our Inner Zombie

Reality... It is eternal because its completeness
and perfection are unrelated to time ...
—W. Somerset Maugham[1]

First off, you need to understand what a 'philosopher's zombie' actually is. There has been a long-running debate in philosophy about the origin of consciousness,[2] and more recently in the cognitive sciences as to whether a computer of sufficient sophistication—one that exactly mimics the capability of the human brain—would, or would not, possess consciousness. Those who answer in the negative argue that if we had the ability to perfectly replicate the human brain in exact detail and allow it to animate an android, the android would act and respond exactly like a conscious human except that there would be no one home. The android would *not* be conscious, it would be 'dark' inside. The android would have no inner psychic life, thus it would be a zombie. This argument has been put forth as a rebuttal to any mechanistic-materialist explanation of consciousness. Those who support this argument can't really say why or how the actual human brain that has consciousness is different from an exact cybernetic replica of the brain. Their point of view is a repackaging of the much older argument that appeals to the idea of a 'soul', or 'emergent properties', or some other inexplicable 'life force' that animates human consciousness.[3]

The zombie of the cognitive sciences is basically a thought problem: if an android could be perfectly programmed to exhibit behavior indistinguishable from that of a human, would there be in this android the psychic phenomena of internal awareness and self-reflection? Or would it be a zombie? To answer this question: there is no logical reason why an android could not be programmed in such a way that it would think that it was a *self,* and that this *self* existed internally and separate from a world that is external to the *self.* This would give the android the sense of separation from its own nature and thus a very human trait. Further, there is no reason why any ideological belief (religious, political, or economic) could not also be programmed into the behavior of the android, along with a strong emotional attachment to these same programmed beliefs.

Then, to turn on the light, and make the android into a true cyborg (a cybernetic organism), the android would be programmed exactly as our culture programs us to be and so the cyborg would also fall into time, and exhibit all the temporal attachments that come with time. The zombie is now brightly lit inside, and possesses a mind filled with concerns of past and future. So, to make this internal "thinking" into self-consciousness, the cyborg would be given an ego by accepting the *memory theory of identity*, which establishes the content of memory as the self-ego. Add to this the primordial, cultural assumption that *memory is past*, and the cyborg becomes a full-blown self-ego that lives and acts in time and space. This zombie is a different beast from the automaton that is 'dark' inside; worse, it is a brain-eating zombie that has swallowed memory.

The cyborg, in order to act and respond just like a human must also have the five bodily senses. And, like the human it will have the sixth sense of memory. But to act like a human, the sixth sense of memory must be split off from the other five senses. It is this schism of the senses that creates the temporal ego, which lodges the inner voice into our mind—the voice which dominates our consciousness. And the root of all this is due to wrongfully thinking that the sixth sense is a historical representation of the ego. The sixth sense has been appropriated by ego.

This cyborg is a zombie that is not so different from a human—it is manifesting all the same dangerous prejudices and misconceptions of a human. Like the human, the cyborg has been taught a confused way of thinking that distorts its reality. This hypothetical android has become like the human who is suffering from the self-other duality. Thinking its ideology has bestowed knowledge, and then reasoning from this knowledge, the android thinks that action could and should be taken. The android then manifests greed, anger, and ignorance, reacting just like humans when their illusions come into conflict with reality. In fact, it is by infecting the android with the human cultural virus that changes it into the conscious cyborg zombie. This zombie-hood not only infects the cyborg, it is the same virus that infects the living human brain.

The cybernetic memory and the human brain are no different in their susceptibility to a cybernetic-type virus. Both human memory and computer memory can host a viral parasite that assumes control and frames the worldview. Human or cyborg, this is a fully conscious zombie, but completely delusional because the mind is controlled by a parasite, a type of virus, not unlike a computer virus. But, if we can understand the nature of the infecting virus and figure out how it was encoded into memory, along with the content of the encoded instructions, would it not be possible that the zombies in question could be restored to reality, by understanding the nature of the madness in their brains?

Consciousness is controlled by a complex of memes that constitutes the self-ego, thus the ego behaves as a self, situated in time and space. Ego can then take action to change the future, responding out of a sense of greed, fear, anger, and ignorance. Buddhism refers to this sort of zombie as a 'hungry ghost'[4], and these 'hungry ghosts' represent most of the world's human population. Philosopher Jacob Needleman speaking of this Buddhist-type zombie said, "The 'hungry ghosts' are starved for 'more' time ... And I understand that it is not exactly more time, more days and years, that we are starved for, it is *the present moment*. Through our increasing absorption in busyness, we have lost the present moment. 'Right away' is not *now*. What a toxic illusion!"[5]

159

"Your life is death already, though you live
And though you see, except half your time
You waste in sleep, and the other half you snore
With eyes wide open, forever seeing dreams ... "
(Lucretius, c. 99–55 BCE)[6]

Lucretius could just as well be describing the 'hungry ghost' of Buddhism. In Western literature the 'zombie' metaphor is sometimes used to denote a very similar condition of *being* in humans—walking comatose through life. The cognitive sciences have shone a bright light onto this phenomenon of sleep-walker/hungry-ghost/zombie. Cognitive science defines a zombie as that which is 'dark inside' (a complete psychic absence), but this is a fictional zombie. If there are to be zombies they will be living, thinking zombies. This is a far more dangerous apparition because this zombie is actually lit-up inside *with an ideology*, and this is far scarier than the walking mindless dead. It is the ideology that makes, and animates, the zombie. The ideology is the program that directs the zombie behavior. Daniel Dennett has explained that if we want to understand the human mind, the thing we need to know is the software program, because the problem with mind is not a hardware problem.[7] Understanding mind is understanding the program that is running on memory, and in fact, it is possible to fully disclose the code that forms the *first principle*, which is nothing other than the cultural supposition that *memory-is-past*.

The thesis that a meme (*memory-is-past*) has made us all into zombies is logically sound, but it is just too hard for most everyone to accept. However, the rational argument and the empirical data strongly support the *self-conscious-zombie* conclusion. The short explanation is that memory has been usurped by a cultural program that is not hard-wired—this isn't in the DNA, rather, it is culturally programmed into the brain. Tadeusz Zawidzki explains:

"Given the importance of cultural learning in human evolution,
our minds, unlike those of other animals, are largely products of

culture. This fact leads Dennett and others to defend a specific model of cultural evolution: the memetic model. The idea first proposed by Dawkins (1976), is that the ideas passed down through culture, called 'memes', behave much like genes passed down through biological reproduction."[8]

Memes permeate and modify our thoughts, and at the very center of this memetic nexus is a central meme, which is a core instruction that dictates: *memory is past.* This is the supposition that turns us all into zombies. By becoming enchanted with the promises of the future, we become, all at once, ignorant, greedy, and angry. To follow the path of time is to follow despair into its maw. The only possible salvation is timelessness.

The ego-conscious zombie is what is eating at the brains of all ideologues. This is a sort of mad zombie, which adopts a cultural collection of memes (often at odds with reality), and thus the zombie is compelled by its faulty belief-system to act as if it knew what it was doing. In this instance the zombie is lit with consciousness but the mind has been taken over by the fundamental meme that there is 'time in the world.' The zombie is conscious, but the mind has been occupied by a viral meme which causes the consciousness to manifest in the form of a mad obsession about *future* and *past* concerns.

The unexamined life is identical to the life of the zombie. As Socrates said: "The unexamined life is a life not worth living." We don't have to imagine a world populated by zombies—the world is already populated by zombies. Philosophy is perhaps the only counter to this. Philosophy looks back to the beginning, to that point where and when we became zombies. The greatest empowerment is doubt—without it 'zombies are forever'. Belief is this strange realm where the door is closed and locked behind you, and then you forget that you still hold the key to the lock, so you can't get out. You are not even allowed to think that this present situation might be delusional.[9] Huang Po, explaining in the vernacular of a 9th century Chinese Zen master, said that humans are trapped in a type of thinking that turns them into zombies: "But while you remain lost in attachments, you condemn your bodies to be lifeless corpses inhabited by demons!"[10]

The zombie condition is caused by a virus for which there is a cure. The virus comes from a universal cultural programming error that fills our head with the insatiable desires of the temporal world. The diagnosis is simple but the cure is enormously difficult. The cure is a philosophical insight which will be resisted and rejected by nearly all who come upon it—and this has been the case for at least the last 2,500 years. The cure is the skeptical path of radical doubt, a path that most are unwilling to take.

The ego that comes to understand itself as just a program running on the human hard-drive, is an ego that knows it is just a form of viral parasite that deforms human nature. This makes us a species of mammal that harbors a big brain, which in turn harbors a parasitic virus that controls the organism's interpretation of the world so that the virus can be propagated to other minds. The meme is a viral plague that infects nearly every human. The virus doesn't kill its host, but takes control of our mind, and so our intent becomes the intent of the virus.

Zombies are created by taking on the belief structure that manifests as an ego in time and space. Anyone who has fallen under the control of an ideology takes on a quality of a zombie. The brain has been taken over by a virus. The virus spells out the presumptions, which in turn frame the religious, political, or economic ideology. The mind is lit up by its ideological certainty, and *Being* becomes subservient to ideology. The zombies are the 'hungry ghosts' of Buddhism, they think of the 'past, present, and future', and this incites greed, ignorance, and anger to flourish—this is the consequence of the fall into time. Perhaps it is so simple that complicated minds can't figure it out. As long as thinking splits off the body's senses from the sense of memory (thinking that the sensations of the five bodily senses are external, and that the psychic sensations are internal), then the mind-body duality will continue to exist.

The program that has infected the mind is such a successful virus in large part because of its simplicity. Just a single supposition, deeply embedded in a cultural tradition that forbids even superficial questioning, because the virus surrounds itself with a protective coat of beliefs. Daniel Dennett explains:

> "People of all faiths have been taught that any such questioning is somehow insulting or demeaning to their faith, and must be an attempt to ridicule their views. What a fine protective screen this virus provides — permitting it to shed antibodies of skepticism effortlessly!"[11]

Culture only endures because it possesses survival value. Culture is the means by which a tradition is conveyed from one generation to the next, and so any meme that can hitch a ride on a culture survives as long as the culture is propagated. However a dangerous meme can also be the end of a culture.

Perhaps the greatest difference between living temporally and living timelessly is that in time there is "action," whereas timelessly, there is only "doing." The first is 'ends' driven, whereas the latter is simply doing what needs to be done. The most nefarious part of living temporally is that there actually is a demon that inhabits your brain and controls your attitudes, emotions, and actions. Anyone who is caught in this situation is a zombie—the meme holds complete control over their thinking. They are mentally controlled by an incubus, a type of cybernetic virus that inhabits the memory, which then conjures up an egoic self that persists in time and space. The ego is the zombie.

The Truth, for anyone who looks, is the understanding that they have been dreaming restlessly while asleep in the Garden. Buddha (The Awakened)[12] is a metaphor for 'waking up' and finding our self back in the Garden that we never left.

Chapter Twenty-one

Strange Loops, Dualities, and Memes

In the ultimate truth there is neither past nor present nor future.
—Nagarjuna, 2nd century CE[1]

The debate about the reality of time has been argued across many disciplines. Though there are those in every discipline who think time is an outright fiction, the great majority either accept time as real or feel that, real or not, it is the way that we see the world and there is nothing that can be done about it.

Some, like philosopher Roberto Unger, are certain of time's reality. He said "nothing is more real than time,"[2] yet, he recognizes the logic of the reasoned arguments against time: "Our mathematical and logical reasoning perpetually suggest to us the reality of a timeless world."[3] He rebuts these reasoned arguments by asserting the presence of an "antitemporal element in our consciousness—the element represented in mathematics and logic."[4] This is an awkward argument for a philosopher to adopt—assuming that both logic and mathematics are contaminated with an 'antitemporal element' and offering only the explanation (not meant literally) that there is a "conspiracy".[5] Unger's argument comes down to the fear that "If time were unreal, however, nothing in our situation would be what it seems to be ... Our lives would be tunnels of illusion from which we could escape, as the perennial philosophy recommends, only by identifying with a timeless, hidden reality."[6] He believes very strongly in time, but his argument against timelessness is out of a fear of a timeless reality. This is definitely an argument that is put forth by his ego, an ego whose own existence is at stake. If you examine time scientifically, logically, and philosophically, time is an illusion, but if you reason the way Unger does, then time is a reality.

Unger is an example of the true believer, but there are many who have looked closely at the evidence, understood and agreed with the timeless argument, and then concluded that the temporal narrative of reality is what we are stuck with. It is argued that it is an unavoidable fact that we experience a time-like world. This is much like the thinking of Immanuel Kant (1724–1804)—he thought this is just the way we perceive the world. As such we don't have access to the actual world that stands behind our perceptions; there is nothing further, philosophically, that can be done.[7] Kant justified our knowledge of time and space by reasoning it to be *a priori* knowledge, meaning prior to or independent of experience. This is a position taken by many who can offer no empirical explanation of time, they just assume that the temporal realm is as close to reality as can be attained, and so they move on.

Philosopher Douglas Hofstadter offers a thoughtful and interesting argument that contends our experience of the world is a myth, but it's what we've got and we are stuck with it. Though Hofstadter has not directly focused on the philosophy of time, he has indicated that our entire conceptual scheme, including time, is a myth.[8] He has spent decades investigating a phenomenon of consciousness that he refers to as "strange loops". He has concluded that our own being, and our own consciousness arises directly out of these strange loops. These strange loops are the source of the self-ego, which he refers to as the "I". He asserts that consciousness itself appears as a consequence of these loops. Hofstadter understands that the "I" is a mental fabrication, but he is certain we are trapped in a loop formed by our way of thinking and there is nothing that can be done about it.[9] He is correct in thinking we are trapped in our habitual way of thinking, but he is mistaken in concluding we can do nothing about it. To fully understand how we landed in this existential cul-de-sac, an examination of Hofstadter's thoughts is helpful.

It is instructive to review Hofstadter's work, because his thoughts provide some further insight into consciousness and ego formation. Hofstadter's first book, published in 1979 was *Gödel, Escher, Bach: An Eternal Golden Braid* (hereafter referred to as *GEB*). The title is a reference to the fact that Hofstadter had observed the 'strange loops' in the musical arrangements of J. S. Bach, in the odd paradoxical illustrations of M. C. Escher's art, and particularly in mathematics of the *incompleteness theorems* of Kurt Gödel.

Hofstadter's *GEB* was enormously successful, winning the Pulitzer Prize for general nonfiction, but he felt that no one grasped the importance of what he was saying. "It sometimes feels as if I had shouted a deeply cherished message out into an empty chasm and nobody heard me."[10] Twenty years later, for the anniversary edition of *GEB*, he wrote a 23-page preface to try again to communicate what was not being heard. And then, 8 years later (2007) there is another attempt with the publication of *I Am A Strange Loop*. Hofstadter is convinced that he has uncovered a fundamental property of consciousness, which can generally be referred to as a 'self-referential theory of consciousness.'[11]

First off, what is a 'strange loop'? Hofstadter has a multitude of examples but no single definition, and many of his examples seem only to be metaphors for the types of strange loops that he envisions give rise to human consciousness, along with the creation of the "I". Some simple metaphors he uses are: two mirrors facing each other, which produces an infinite regression of reflections that diminish in size toward a vanishing point, or similarly, pointing a TV camera toward a TV screen that is simultaneously displaying the image that the camera is recording, thus producing a video feedback loop, or a simple microphone-amplifier feed back, but he is explicit in saying that these simple feedback loops do not possess the 'strange' quality.[12] As metaphors these loops suggest self-reflection and looking back, or into, the experience that is being experienced.

Hofstadter's examples in the graphic arts, music, and mathematics, are loops that are frequently more complicated. The characteristic that gives the loop the quality of 'strange', according to Hofstadter, is the presence of symbols. "It is the upward leap from *raw stimuli* to *symbols* that imbues the loop with 'strangeness'."[13] The consequence of the great proliferation of these loops in our experience of the world results in the emergence of self-consciousness. Hofstadter later said, "Consciousness is the dance of symbols inside the cranium. Or, to make it even more pithy, consciousness is thinking."[14]

Alas, the conclusion of *I Am A Strange Loop* leaves Hofstadter still grappling with explicating the phenomena that he has so intently studied. He cannot explain the first person experience that is characteristic of the 'strange loop' which results in the awareness of the "I".[15] All that he is able to offer is a description of the phenomena, but no matter how good the description, it doesn't serve as an explanation of the leap into the conscious "I". Hofstadter never actually explains how the "I" is spun out of these loops other than to suggest that it is an emergence out of complexity. He said about the "I", "The notion that such a pattern grows enormously in size and complexity over time, perceives itself, and entrenches itself so deeply as to become all but undislodgeable will constitute a satisfactory answer for some seekers of the truth ... For others, however ... it will not do at all."[16] This is the answer Hofstadter leaves us with, but there is much more to say about the strangeness of the loop that creates the sense of "I".

Hofstadter emphasizes the *simple* feedback-loops but understands that the human organism is made up of a myriad of feedback-loops which together "grow enormously in size and complexity over time." Indeed, the human body is a myriad of feedback-loops that control temperature, blood chemistry, body hydration, heart rate, etc., altogether referred to as homeostasis—a 'striving' for equilibrium. These are physical/chemical processes that are 'seeking' equilibrium, however the mental processes that give rise to an "I" are, as Hofstadter said, embodied in our symbols and our thinking. Now we are talking about language, and it is the understanding of language that will unravel the strangeness of the loop that is the "I". Keep in mind that talking, writing, thinking takes place in language and that words are nothing other than symbols, and even the symbols of a mathematical system constitute a language.

There is a common property of all languages: they are all self-referential. Every word of a language refers to other words in the language and is defined by other words within the language. Words express a difference from other words; this is what gives them meaning. A language is a self-referential loop and it forms a hermeneutic circle.[17] It is clear that the *strange loops* Hofstadter has written exhaustively about are, by another name, *hermeneutic circles*.

There is something more that needs to be said about this strange loop that creates an "I" out of symbols and thinking. It is undoubtedly a *hermeneutic circle*, but there are some constituent components that are important to the self-referential workings of the circle. These components are the many dualities of language allowing us to express the range of perceived and conceptual differences. The most obvious are the dualities formed by two words that are polar opposites such as 'hot and cold' or 'good and evil'. Often there is also an array of words that allow the expression of increments that lie between the polar extremes—we have invented all kinds of modifiers to describe intermediate positions between the opposite poles. Overall, there might be hundreds or maybe thousands of these dualities scattered throughout language which allow the speaker to express fine discriminations in perceived phenomena, or to speak of the nuances of various concepts. Always the meaning is given as much by the word spoken as by the contrasting expression that was not spoken.[18] One side of a duality always reflects the other, and together they form a small part of the overall hermeneutic circle that constitutes the language we speak.

We are brought back to Hofstadter's 'strange loop.' Indeed, the loop is made of thinking and symbols, but it is clear the loop is also a hermeneutic circle and the symbols comprising the loop are the words of our language. The composition of mathematical formulae, or the composition of music, or the composition of sentences, are all composed from the hermeneutic of our symbolic language.[19] Still, the above explanation does not specify how the "I" of the self-ego emerges out of the hermeneutic. However, there is a particular species of duality that is markedly different from all other dualistic oppositions, and it is here that we find a key ingredient of "I". This 'strange' species of duality takes the general form of *belief/doubt*. We all carry around in our head an abundance of this sort of duality.

The general belief/doubt duality, which is so instrumental in how we think, is represented by belief on one side, and doubt on the other. The belief side of the belief/doubt duality is responsible for all the ideologies that humans fabricate, along with all other 'knowledge' for which we have no evidence.

The most curious, and the most insidious aspect of a belief is the way in which the belief, through the act of believing, goes into hiding from our consciousness. As long as we are conscious of a belief as being a belief, we are cognizant that the belief entails a degree of uncertainty. The fact that it is a belief requires it be paired with some level of doubt. However, as a belief moves along the belief/doubt spectrum, ever closer to complete-belief, the doubt slips from consciousness until the idea (the belief) becomes an article of unquestioning faith. Oddly, what was once belief, now appears as truth. There is also another equally interesting aspect of this belief/doubt spectrum. At the 'complete doubt' end of the spectrum, where there is the manifestation of complete-doubt, both belief and doubt completely disappear. Doubt never spontaneously appears, it is always triggered by the presence of belief. This reveals that any belief/doubt duality always originates from the act of believing. When there is complete doubt, even doubt disappears, since there is nothing further to doubt. On the other hand, when there is the manifestation of complete belief, belief doesn't evaporate, it just slips over into faith, and the fact that it is only a belief, is a fact that becomes hidden from the believer. As Richard Dawkins puts it, "Indeed the fact that true faith doesn't need evidence is held up as its greatest virtue ..."[20] Dawkins explains further, "Faith is such a successful brainwasher in its own favor, especially a brainwasher of children, that it is hard to break its hold. But what, after all, is faith? It is a state of mind that leads people to believe something—it doesn't matter what—in the total absence of supporting evidence."[21]

Dawkins came up with an intriguing and original explanation of how many of our beliefs are instilled into the way we think. The theory and science that his idea spawned has come to be called *memetics*.[22] The theory of memetics is a powerful psychological tool that explains a fair amount of why we think as we do, and postulates a theory of how the self is structured. Enabled by the theory of memetics we are given a way to examine and analyze how the beliefs and how the "I" come to inhabit consciousness. The study of memetics reveals a theory of mind that proposes a new psychology, and provides a mechanism to explain cultural evolution.[23]

A specific aspect of memetics that is of particular importance to *Time Sutra* is the idea that certain of the memes[24] inhabiting our brain possess a viral-like nature that is overall detrimental to the human that is 'infected' with these particular viral memes. Simply put, this occurs because the brain is similar enough in basic function to that of a modern computer, such that the brain is vulnerable to mental viruses which are propagated much like computer viruses.[25] Said in another way, the primary function of our brain is to serve as a memory, and there is no legitimate reason to think that it is so different in structure and function from the memory of a modern digital computer that it would not be susceptible to infection from a cybernetic-like virus.[26] These 'mental viruses' are a type of meme, but also keep in mind that "The vast majority of memes are not viruses but are the very foundation of our lives."[27] However, there are the dangerous memes, and in one way or another, the danger derives from the fact that many memes are beliefs that are detached from reality. And the stronger the belief the greater the potential danger. Of special concern is a single viral meme that splits off the sixth sense from the rest of the five bodily senses, and so divides our world into the experience of the mind/body duality. Through faith that *memory is past*, this most harmful meme splits the world into self and other.

In summary, Douglas Hofstadter's analysis of 'strange loops' is an innovative investigation into the emergence of the self-ego—which he refers to as the "I". In so many ways his description of the strange loops very closely parallel what might elsewhere be called hermeneutic circles. Emerging out of all the self-referential symbols that constitute our language, the "I" is brought to a self-conscious thinking. The type of thinking that humans generally engage in is greatly animated by *what they think they know*. But, the knowledge is invariably populated with beliefs that are viewed as facts, even though the beliefs lack any connection to empirical experience. How we come to imbibe all these beliefs often happens unknowingly through immersion in our cultural tradition. The thesis of *Time Sutra* is that there exists in all humans a single rudimentary belief that conditions our thinking, and does so at a very early age. We are wise to think of this belief as a viral meme, a meme that our culture unknowingly inflicts on us. The meme can be expressed simply as *memory is past*, but the idea is distributed globally throughout the brain, such that, at any point of contact with phenomena of memory, our experience of these phenomena passes through the *'memory is past'* filter, providing us with an artificially constructed temporal view of the world. We are alienated from the atemporal world that is our nature.

Remember, by thinking that *memory is past*, we split the six senses into body and mind. The sixth sense comes to represent what is going on in the mind of the self, whereas the other five senses represent what is going on in the body. This produces the sensation of the body carrying

172

around a brain that contains the "I", which is experiencing the external world through the bodily senses. Prior to thinking *memory is past*, the six senses were integrated and there was no differentiation between mind and body. There was no opposition between the thinking "I" and the sensory body, no distinction between interior thought and exterior experience, because it didn't exist. A single belief turns the world into an illusion that is so compelling, and at the same time so confusing, that escape back into reality is nearly impossible. We never find our way out of the belief, unless we are able to abandon our ego. Only a Buddha awakens from this dream.

Part VII

Resolving Time

Chapter Twenty-two

The Western Buddha

Applicants for wisdom, do what I have done: Inquire within.
—Heraclitus, 6th century BCE [1]

Buddhist philosophy is deeply involved in questions about time, self-ego, mind and memory. A closer look at this philosophy is very illuminating because Buddhism has something to say about a modern perennial philosophy just as modernity has something to say about what a modern Buddhism will be. The following is a brief explanation of this ancient philosophy and how it might resolve some of the problems of modernity.

Buddhism enters a culture, and over time it modifies itself and adapts to its new culture by taking on the character of the host culture. The historical record shows that this assimilation often takes place over a period of a few hundred years. "Wherever it was introduced, Buddhism's initial appeal was always to the social elite. The merchants, nobility, intellectuals, and even the warriors were the first to adopt Buddhism, and over the years it usually filtered down to the masses."[2] Slowly the flavor of the host culture becomes incorporated into a new expression of the traditional teaching. This same pattern of absorption of Buddhism into a new culture has been repeated throughout the history of the tradition. Buddhism never imposes itself on a culture, instead it becomes infused with the culture and blends into the cultural character. "Mu Soeng points out that as Buddhism enters new countries and cultures it subtly incorporates and adapts preexisting beliefs and practices."[3] The tradition inevitably is modified to mirror the culture, and thus becomes that particular culture's expression of the Buddha's teaching.

177

Knowing how Buddhism evolves within a culture has, over the past few decades, stimulated frequent speculation in the Western Buddhist community about the eventual nature of mature Western Buddhism. What will this newly emerging Buddhism ultimately look like? What will be the Western cultural values and characteristics that this modern Buddhism will assimilate? Some of this speculation is specifically about American Buddhism, or European Buddhism, or more generally about Modern Buddhism.[4] But there is good reason to call this new practice *Western Buddhism*, since whatever form contemporary Buddhism eventually assumes, it will essentially be different in character because it becomes infused with Western culture. Knowledge of the tradition in both Europe and America has been around for well over a hundred years, but was confined to a small subculture of practicing immigrants and interested academics. It has been practiced in America at least since the time of the immigration of the Chinese in the 19th century, but was generally isolated from the broader culture.

Sociologist James Coleman captures the post-war situation: "By the end of World War II, Buddhism had made only the shallowest penetration of Western culture."[5] Buddhism was never assimilated into American culture, and never attained a distinctive American flavor until the advent and subsequent influence of the Beat Generation of the late 1940s through the 50s. The Beats are now considered by most to have been instrumental in introducing Buddhism into popular American culture.[6] Coleman explains, "The Beat's vehement rejection of conventional American culture opened the door to new perspectives on life, and it was among this group that Buddhism found its first broad appeal in the West."[7] It was the writers and poets, namely, Jack Kerouac, Allen Ginsberg, Gary Snyder, among others, who grabbed the attention of popular culture. Author and practicing Buddhist Melvin McLeod recently wrote about this influence:

"Buddhism began to have a real impact on American culture in the 1950s, with Zen's influence on the Beat writers and other artists and intellectuals. So we are only six decades into what might be a two-hundred-year process to establish a Western Buddhism that is genuine and complete. Some say that Buddhism is still in its infancy in the West, but I think perhaps it is a little more advanced than that, a toddler now taking its first hesitant steps on its own."[8]

The generation that followed the Beats, those coming of age in the late 1950s and early 60s, having been exposed to the Beat literature, felt an affinity for Buddhism—it was during this time that the philosophy truly infected the American culture. The cultural changes of the 1960s and 70s, and in particular the psychedelic culture, created an atmosphere that was generally accepting of Eastern practices of Yoga, Vedanta, and Buddhism. "It is undeniable that a significant proportion of those drawn to Buddhism and other Eastern traditions in the 1960s ... were influenced in their choice of religious orientation by experiences induced by psychoactive substances such as marijuana and LSD."[9] At present Buddhist practice remains a small subculture in America, but it has now been accepted by the mainstream as one of the available religious alternatives that one might choose.

Although the philosophy has begun to take on Western influence, there is no new Western school of Buddhism that has emerged from the many Asian lineages now being taught in the West. There are however some distinct adaptations that are noticeable. Western practitioners are less monastic and more lay oriented. The Buddhist communities are less hierarchical and paternalistic than their Asian counterparts from which they derive. Teaching positions are populated with more women than was customary in the East, and this is probably due to the fact that the practices in most Western Buddhist communities are not segregated by gender. These changes are necessary for there to be a Western Buddhism and are essential to the flavor of a modern Buddhism, but this is all superficial in comparison to the Western philosophical insight that will define the fully developed Western Buddhism.

The question remains, what will the practice come to look like in its mature Western incarnation? Fortunately, this can be predicted with a fair degree of accuracy by simply understanding how Buddhism operates and knowing the unalterable core philosophy that lies at the heart of Buddhist enlightenment. First, it is important to know how Buddhism penetrates a foreign culture. Because Buddhism doesn't proselytize, it is most likely that Buddhist teachings are something that a foreign traveler brings home to their own culture. Teachers have been sent out from Asia to teach their respective practices, but this is generally done because there is an audience already present. There is a very good reason why Buddhists don't proselytize—mostly because Buddhism claims no relative or conceptual knowledge, so there are no beliefs to preach. The Buddhist who proselytizes does not know Buddhism. Even if Buddhism penetrates the foreign culture through immigration, the introduced Buddhism generally stays confined to the immigrant community, and unless one goes looking, you won't find Buddhism. It is the nature of Buddhism that only those who go looking ever have a chance of comprehending Buddhism.

Buddhism's presence in a foreign culture is unassuming and non-threatening to the established norms of the society. Unlike other rival religions or ideologies, the local customs are not challenged. Buddhism has no competing god or religious beliefs to promote. Buddhism surreptitiously slips into culture. It bypasses the host culture's first line of defense by making no pronouncements of truth or authority. Buddhism offers no abiding paradigm, thus it seems to be of no threat to the host culture. Cultures normally defend their turf by condemning any alternative cultural beliefs, but with Buddhism there are no beliefs to give offense. This is not a tactic, this is the nature of Buddhism.

Buddhist philosophy has been widely recognized as a logical argument that takes the form of a deconstruction. The term *deconstruction* has in recent decades been closely associated with postmodernism and linguistic theory—it was originally derived from

Martin Heidegger's *destruktion*. However, the theory of deconstruction as employed by Jacques Derrida and others is often involved in complicated linguistic analysis, but the Buddhist meaning and use of the term *deconstruction* is direct and uncomplicated. Essentially, for Buddhism, it means disassembling, or taking apart, the conceptual scheme that forms the ground of one's worldview. Our individual conceptual framework forms the way we see the world. In fact, this conceptual framework is our 'personal philosophy' (whether we acknowledge it, or are even aware of it). Like all philosophies, our personal philosophy is grounded on some fundamental assumptions. Built atop these basic assumptions are further suppositions. In formal philosophy these structural components are variously named premises, axioms, principles, etc. Thus, for Buddhism a deconstruction is a careful examination intended to identify these structural components and pick them apart, with the motive of disassembling the entire mental edifice that causes us to think as if we are a self-ego living in a spatio-temporal world.

Modern writers on Buddhism frequently cite the deconstructive nature of Buddhist practice and philosophy. Buddhist scholar Mu Soeng said, "When Buddhist teachings are practiced authentically, there's no choice but to deconstruct the inherited psychic structures."[10] *A Sourcebook in Indian Philosophy* speaks of the deconstructionist method of Buddhism stating, "For the great Buddhist thinkers, logic was the main arsenal where were forged the weapons of universal destructive criticism."[11] Professor of Japanese religions at Harvard University Ryuichi Abe remarks that "The Mahayana theory of nonduality demonstrates a striking parallel to contemporary deconstructionist theories."[12] James William Coleman said of Buddhism, "Its most profound teachings do not offer a new identity or a new set of techniques for managing our problems but deconstruct the whole project of the self—to bring us to see the pointlessness of our desperate efforts to construct, maintain, and protect our self-identity that consume so much of our lives."[13] Buddhist philosopher David Loy in his essay *The Mahayana Deconstruction of Time*, comes closest to revealing that our great mistake is in thinking *memory is past*. He says:

"So Vedanta and Buddhism both emphasize the role of memory 'wrongly interpreted': identifying with such memories provides the illusion of continuity—a 'life history'—necessary to sustain a reified sense-of-self. Thus past and future originate and work together to obscure the present, usually negating it so successfully that we can hardly be said to experience it ..."[14]

What, is it that Buddhism is supposed to deconstruct that will make it uniquely a Western Buddhism? As in every culture, the Buddhist deconstruction is a dismantlement of ego, a rejection of the self-other opposition, and the elimination of time. The Western tradition will be distinctive because it has four hundred years of empirical observation of the very things with which Buddhism is concerned. These observations, which have become increasingly precise over time, are made and interpreted within the context of the spacetime continuum. The spacetime continuum, in the thinking of Thomas Kuhn is called the 'ruling' paradigm.[15] Generally speaking the ruling paradigm is made up of widely accepted suppositions that form the worldview through which a culture experiences that which is thought to be 'reality'. Kuhn, though this was not his objective, defines exactly what Buddhism wants to deconstruct: the ruling paradigm.[16]

What Kuhn means by the ruling paradigm is the intellectual and scientific construct that the great majority of scientists and intellectuals currently subscribe to. The ruling paradigm changes over time, and Kuhn calls the more radical changes, *paradigm shifts*. The prime example of this type of paradigm shift taking place in our conceptual worldview is the profound difference in thinking that occurs when the believer in an 'Earth-centered universe' shifts perspective to believing in a 'sun-centered universe'. Suddenly, the old religion doesn't work as well as it used to. The Buddhist deconstruction, however, is an even more radical paradigm shift which results from understanding that there can be no center, because there is neither time nor space in which to place a center. And there is no paradigm shift, because there is no paradigm remaining that can replace the deconstruction—all that remains is our original nature.

The curriculum at universities throughout the world is the teaching of the Western paradigm. Even in countries in the grip of fundamentalist religion, the demonstrated technical power of the modern paradigm cannot be ignored. To even participate in the modern world requires the acceptance of large amounts of technology that has been spun out of empirical science. Never in world history has there been this broad acceptance of a single cultural paradigm, and this presents Buddhism with an opportunity. The precisely articulated paradigm of the Western tradition contains, inherent within its structure, an equally precisely articulated Western Buddhist deconstruction. This offers Buddhism the possibility of what is perhaps the clearest dharma[17] ever spoken. Buddhism's greatest problem throughout its 2,500 year history has been the failure of transmission of wisdom from one human to another.

The lineages of successive teachers and students are eventually corrupted by time, and the profound insight is lost.[18] As an example of this decay the Zen tradition has many times had sparks of brilliance, but various lineages often degrade into a very questionable form of communication, or non-communication. Zen teachers will sometimes answer a student's question with a shout, slap, or silence.[19] Those who claim to be teachers seem unable to teach what they claim to know. The transmission of wisdom is impeded by the lack of a clearly spoken dharma. However, Western Buddhism will have the opportunity to devise a dharma-teaching that can be understood by a large segment of the population throughout the world's various cultures.

Modernity has produced the most detailed explanation of time that any culture has ever achieved, and the paradigm has gained worldwide acceptance. The most educated and influential people worldwide have been exposed to the Western paradigm, and have accepted many of the conclusions and consequences which come with the adoption of modern science.

The Western treatment of *time* is far more precise than any other temporal analysis throughout world history. We have measured time with the greatest precision; we have studied, in detail, the psychology of time; we have 2,500 years of philosophizing about time; and physicists, for almost a hundred years, have had increasing doubts about the efficacy of time. The Western paradigm that a mature Western Buddhism would deconstruct is the impressive structure that modernity has built, which is none other than the modern scientific worldview, the temporal continuum in space and time. A Western Buddhist deconstruction will track down that most elementary supposition that gives rise to temporal thinking, and rip it out. The one who can perform this feat is the Western Buddha.

The Buddhist deconstruction becomes, in a strange way, the completion of modernity. After all, the intent of any legitimate philosophy is truth, and modernity launched an investigation based on sense and reason, confirmed by observation, which often resulted in the shattering of many cultural icons. Everything for all of time was captured in the cause-effect, deterministic explanation, provided by the modern scientific spacetime continuum. And the evidence of verifiable empirical measurement guaranteed the integrity of that paradigm. The one insurmountable problem, which is impossible to avoid, is the fact that there is no empirical evidence for time! Nevertheless, this problem has been suppressed, ignored, disregarded, overlooked, and otherwise evaded. Our misunderstanding about time is a human affliction that comes out of the cultural indoctrination, and it is an indoctrination that starts in infancy. Culture may not be able to divest itself of temporality, but if it can, it will be through philosophy—and that philosophy will look a lot like Buddhism.[20] A Buddhist deconstruction, by insisting on observational verification of time, is only carrying the logic of modernity to its natural conclusion, the conclusion that being is timeless.

The mature Western Buddhism describes the disease, explains how the contamination occurs, and prescribes a remedy.[21] When you already have the disease, and we all do, it is nearly impossible to convince 'the diseased' that they are infected, when the main strategy of the infecting virus is to hide its own presence. Reality is not hiding itself, rather, it is the *virus* doing its best to hide reality. The virus grabs memory and occupies it as its own, and imposes a temporal interpretation that essentially cedes all experience of the sixth sense to ego. There is only one way that society can ever be redeemed: we must return to the unchanging reality that persisted before the time-virus took over our being.

Chapter Twenty-three

Allegories of the Labyrinth

In a state of ignorance (time) is the first thing to manifest itself,
but in the state of wisdom it disappears.
—Bhartrhari (5th century CE)[1]

Introduction: This chapter is an investigation of four labyrinths. Each of the four labyrinths entraps their victims in confusion. The four labyrinths to be considered are *Plato's Cave*, Jorge Luis Borges' great *Library of Babel*, the ancient memory technique of the *Memory Palace* (constructed out of the sixth sense of memory), and finally the subterranean maze inhabited by the *Minotaur*.

We are trapped in a labyrinth of our own making. Humans have thought their way into their present predicament. Many of us don't even recognize that there is a "present predicament"—and if you don't recognize that there is a problem, then it is impossible to look for a solution. We have forgotten how and why we arrived here in the first place. The labyrinth in which we are trapped is our own mind. We have adopted a way of thinking that creates a dream world. We not only don't remember how we created this dream world, we don't even know it is a dream, so there is no attempt to try to remember.

The labyrinth is our own mind—in particular it is the way in which we have trained ourselves to think, or, more correctly, the way in which our culture has trained us to think. Memory has been usurped by a type of thinking that entails forgetting the nature of original memory.

The labyrinth is a dark and forbidding place. It contains a mystery of the unknown that is experienced as an abiding confusion. The structure of the labyrinth is meant to confuse—if you find your way in, you may

187

not find your way out again. Those who are entrapped are caught in confusion. Importantly, there is a way out of the confusion. If the design of the labyrinth in which we are trapped can be discovered, then the confusion abates. The problem then becomes finding a *clue*, or perhaps a *sutra* that will inform us of the nature of the trap in which we are caught.

Clue and *sutra* are both derived from Sanskrit, and both literally mean thread. *Clew* is a ball of thread, and is the precursor to the term *clue*. *Sutra*, when used to describe the teachings of the Buddha, is a shorthand mnemonic device written to bring the lengthy teachings to memory. Whether clue or sutra, both are the *thread of reason*, and both are intended to extricate ourselves from the delusion that entraps us in the human condition. The ancient Greek labyrinth of the Minotaur is solved with a clue, while the labyrinth of our own mind is solved with a sutra, but both are threads of reason.

The point is that we are trapped in the labyrinthine machinations of our mind—Buddhists would say that we have forgotten the nature of our own mind. In general, there is only one type of key that will unlock the mystery of the labyrinth and free our mind. The solution will always be one of turning the light around, and looking within. We all have this capacity to remember something very important, something that we have forgotten. Socrates felt that we all possessed wisdom—if only we could remember. The solution to the labyrinth is remembering to come to our senses—in particular, remembering the nature of our sixth sense.

Time Sutra is written as a mnemonic clue intended to bring to light our remembrance of reality. It is a mnemonic device that reveals our true memory which has been covered over by the 2,500 years of cultural accretion occurring since the time of the Buddha. And in the West, for 2,500 years, we have been unable to remember what it was that Parmenides revealed in his poem *On Nature*. *Time Sutra* is a reminder of where and when we began thinking the way we do. The key to the labyrinths is finding the *beginning*. This is the solution to time. *Time*

Sutra is the clue. If you can follow the thread of the argument then you will know there is something you have forgotten—and you will know it is the source of our alienation from the natural world.

The Cave Allegory: One of the earliest recorded descriptions of the labyrinth is Plato's 'allegory of the cave'. Plato describes the inhabitants of the cave as chained prisoners, but in reality it is the prisoners' own thoughts that are the constraining chains. The cave itself isn't described as a labyrinth, but the thinking that has its hold over the prisoners in the cave can only be described as a labyrinth. From this maze of confused thoughts, very few prisoners ever escape into the light.

Plato's cave allegory is told in the words of Socrates, who imagines an underground cavern where prisoners are confined from early childhood. The prisoners are so constrained that they can only see the cave wall in front of them, on which are cast shadows that originate from behind the prisoners. Throughout the prisoners' life, there is a parade of shadows displayed on the cave wall and the prisoners come to recognize and name those images that they are familiar with. The prisoners speculate and theorize about the relationships among the shadows. In essence, the world of shadows constitutes reality for these prisoners.

Socrates then describes one of the prisoners who is able to free himself and is then capable of turning his gaze around[2] and looking toward the light. At first, the prisoner is blinded by the light, but slowly comes to understand that his world had been an illusion. The prisoner sees the reality of the light and realizes that solid objects created the shadows and so he understands that his former world was merely a shadow of reality. The prisoner eventually emerges from the cave into the sunlight, finding the real world of infinite color and variety.

Socrates then tells of the consequences should the former prisoner return to the cave to enlighten and free his former companions. The remaining prisoners would not only resist being freed, they would consider the revelation to be madness, and Socrates suggests that if the free man persisted, the other prisoners would kill him if they could.[3]

Plato considered the allegory of the cave to be a portrayal of the state of human ignorance. The human condition is a state of persistent ignorance in which a shadowy illusion is mistaken for reality. Interestingly, Plato attributes this condition to *forgetting*. According to Plato, the act of learning the truth is, in fact, *remembering* what has been forgotten.[4] Plato, and by inference, Socrates, understood that our failure to know Truth is connected to our failure to correctly perceive memory.

There is a very real modern-day analog to Plato's cave, but rather than prisoners in a cave staring at shadows on the wall, affluent modern humans sit in their living rooms and stare at large flat-panel TVs on the wall. These humans are transfixed by the display of video games, movies, Internet, and 500 cable channels. There are no physical restraints, but they cannot turn their gaze away because they are so entertained. Most of the entertainment comes with a constant stream of commercials intended to remind the viewer of what they want, or to encourage the viewer to want something that they didn't previously know they wanted. And what is wanted most of all are the gadgets that deliver the entertainment. Since newer and better gadgets are continually produced, there is an unending renewal of wanting. They must leave their cave to work and shop, so as to pay for the entertainment in their cave, along with keeping their cave warm and supplied with provisions. But, nervous outside the cave, they stare into the portable electronic gadgets that keep them connected, sustaining their need to be entertained, or at least distracted. This all but prevents the introspection and solitude that are generally thought to be the necessary circumstances for 'turning the gaze around', and ultimately allowing escape from the labyrinth of entertaining diversions.

The Library of Babel: Jorges Luis Borges published the *Library of Babel* as a short story in 1944.[5] Throughout much of his writings there is a marked obsession with *identity*, *time* and *labyrinths*. The library allegory, like the 'modern cave allegory', repeats the information-overload that prevents us from seeing the truth, but unlike the unlimited entertainment in the modern cave allegory, the Library of Babel is a near-infinite repository of information. Every truth is contained somewhere in the volumes that line the shelves, but there is no way to know where to start in the search for truth. The search for truth always entails, at first, the search for that point where the search will begin.

Each corridor opens onto a cavernous, hexagonal room lined with bookshelves, with six radiating corridors leading to similar cavernous rooms, and each room radiating six corridors—and this honeycomb-like architecture is repeated for as far as anyone can know. Spiral staircases connect to the floors above and floors below—each floor having the same identical architecture. No one knows or can conceive of what is outside or beyond the library, nor is there a basis to speculate on how long the library has existed.[6]

The bookshelves are filled with texts of similar size, binding, and number of pages. Sometimes the books are filled with all the greatest literary and scientific works that have ever been, or ever will be, written. But mostly the books contain nothing of consequence, and usually books are nothing but gibberish. There was no equivalent of the 'library catalog', although there must exist an equivalent index somewhere on the bookshelves, since all information was contained on the bookshelves. The people that populated this universe were librarians, they had a reverence for information and knowledge, but the nature of the library drove many to suicide.

Analysis of the library, by astute librarians, indicated that all possible written information was contained within the library, and when this understanding was reported to the librarians, there was jubilation—at first.[7] Then frustration, followed by despair, because the Truth was known to be present on the shelves, but it was hidden in the vast amount of irrelevant or even incoherent babble that filled most of the texts. For a time, there was even a band of librarians who set out to destroy all of the volumes of nonsense that filled the shelves. They reasoned that eventually the library could be distilled down to only the books that were meaningful.

Borges' library allegory seems to anticipate the Internet. In fact, had Borges lived a few decades longer he would have experienced his Library of Babel first hand—in the form of the Internet. The near-infinite Internet seems to contain almost all knowledge, but it also contains the ignorance, the trivial and the nonsensical, and these latter constitute the overwhelming content of this contemporary Tower of Babel. Construction of the biblical Tower of Babel failed because God confounded man with so many languages that communication and social coordination became impossible. With the Internet we can understand the languages that are being spoken, but the problem is that everyone is talking and no one is listening.

The library allegory, like the modern cave allegory, emphasizes the sensory overload that prevents any solution to the maze in which we find ourselves. In the instance of the modern cave allegory, we are so completely distracted by entertainment we are not even aware that we are imprisoned in an illusion. The library allegory tells how we are overwhelmed with information[8], and it is information that could lead to the truth, but there is not even a clue as where we should start looking. To begin searching, without knowing where to begin, through the near infinite amount of data, is a strategy that will preserve our ignorance until, inevitably, in time, there is death.[9]

Always, there must be a search for the beginning, along with an understanding of where to first start the search. But the beginning is not in time, because the beginning is always now. Nor is the beginning a place, because the place is always here. However, if you think of here and now as the beginning of something that will manifest in the future, you have not understood the beginning. The beginning is the beginning of time, but this is not the kind of beginning tied to when the clock started ticking, it is a beginning whose onset is at the instant when we *mistake* memory for something that transpired in the past.[10] This is the fall into time.

The Memory Palace: The Memory Palace is a very ancient mnemonic device that was important in the transmission of learning and culture before the age of printing. The technique, also known as the *Art of Memory*,[11] has been around at least since the middle of the 1st millennium BCE. It has been variously attributed to the Greek poet Simonides (c. 556–468 BCE), or the cult-like Pythagoreans, or possibly having its origin in Egypt. *Memory Palace* is defined as "One of the most ancient mnemonic strategies known, in which items are mentally placed in a huge architectural edifice [of the imagination]; to recall the item, a person recalls the location where the objects or words are placed."[12]

There is a surprising parallel between the Memory Palace, which is a conscious mental construction, and our cultural worldview, which is mostly an unconscious construction that is derived, unexamined, from our youthful social indoctrination. In his research, Jean Piaget closely inspected the child's mental development of thinking in terms of time, space, and self. He concluded that the way we conceive time and space is taught through culture. Most importantly, Piaget recognized that there is no fully developed self-concept until the child has a firm grasp of the temporal, and their own place in the temporal world. Piaget explains, "... it is by learning to tell stories to others that a child learns to tell stories to himself and thus to organize his active memory."[13] The child is successfully enculturated into a society when they interpret their own being as something constituted out of their memories, and they learn that memories represent the past self. Through this they are taught to create a memorial construction of the self (much like a Memory Palace), extended in time and space. The child then becomes a self-ego situated, and presumably living, in time and space. The Memory Palace mnemonic technique simply takes advantage of the way we have already learned to think, and then organizes memories in a similar, albeit intentional and conscious, fashion.

By the time we reach adulthood we have had lots of practice locating the self in space and time by telling our personal story to our self and to others. The culturally constructed self-ego already is, in essence, a kind of Memory Palace, where there are corridors penetrating deep into times past. The self is arrayed throughout the area of the self's experience, which is defined as our episodic memory.[14]

The Memory Palace, as it is constructed, is a mnemonic device that mimics the time-space paradigm which our culture has already imprinted on memory. The memories are the memories of our self that are already arrayed in time and space. These are from the *episodic-autobiographical* memories and have been keyed into the time-space framework. The mnemonic technique of the Memory Palace is to allow encoding of semantic memories (memories that are not stamped with spatial or temporal tags) into a spatial organization. Doing this is somewhat intuitive, because this is very similar to how the ego is originally encoded.[15] Ego is built out of the organization of the episodic memory (also called the autobiographical memory), but it then goes on to claim ownership of the semantic memory. Ego employs the mnemonic techniques as a way of enhancing the power of semantic memory. However, the discrimination between episodic and semantic memories is clearly arbitrary. As philosopher M. G. F. Martin said, "If there is no essential difference between episodic and semantic memory, but only a difference in the way information retained is used, either as a story about one's past, or as an account of how the world in general is, then no such memory has a special link to the past. There can be nothing which counts as experience of the past."[16]

The grand edifice of the self—our constructed ego—is, in essence, a *memory palace* built on a very insecure foundation—it is based on a fundamental misperception of our sense of memory. When you identify with the ego you become an inhabitant of the labyrinth that the ego has

194

created. The ego is the Minotaur that you must kill before you can escape the maze.

Ariadne's Thread: The myth of the Labyrinth and the Minotaur, one of the oldest of the labyrinth myths, tells of how the god Theseus descends into the Labyrinth and slays the Minotaur, a creature with the head of a bull and the body of a man. Anyone who had ever entered into the depths of this Labyrinth was forever lost, because they could never escape the confusion of the maze. However, the goddess Ariadne—who was in love with Theseus—spun a thread that Theseus could lay down behind himself on his path as he journeyed into the maze. Thus, Theseus was able, not only to enter into the Labyrinth and kill the Minotaur, but to escape from the maze. Ariadne had supplied the clue that was the solution to the Labyrinth.

We find ourselves at the center of a deep and confusing labyrinth, in complete darkness, having no idea how we arrived here and no idea about the path out. But it is worse than that. We can't even look for the path out of our confusion because we don't realize that we are living in an illusion, so there is no reason to even suspect that reality might be completely different from the way we currently think. And even if we do suspect that a reality exists beyond this dream, we have no way to find the beginning of our search for the path out of the labyrinth and into the light. However, there is a method. There is a *clue,*[17] and it is always available.

At the very center of the labyrinth, in the depths of confusion, sit down in the darkness and quiet the mind. This is an ancient technique that a few have called the Perennial Philosophy,[18] because all of the eternal methods recur naturally and reasonably and converge on a single practice. Simply reach down and touch the philosophical ground, because there on that ground, in the darkness, lies the *thread*. This is the *Thread of Reason*, and it extends all the way back to the light at the beginning. Here at the very center of the labyrinth this *thread* that we pick up is the very distal end of the thread of reason, and it is a long way to the light. We got here by the way we think. So if we are careful, and closely analyze our own memory, we can also realize *how and why* we got here, which in turn, will reveal how we must think our way out of here.[19] This practice is nothing more than remembering what has been forgotten. Re-establish a true contact with the sixth sense—reach down and pick up Ariadne's Thread and follow it to the light.

EPILOGUE

The senses are the beginning and the end of human knowledge.
—Michel de Montaigne 1533–92 C E [1]

We are standing outside the Garden. The path that we walked on, out of the Garden, is now so overgrown that it is enormously difficult to retrace our steps. But the thread is still there and it is the only thread we have. If this thread of reason, logic and observation does not reveal reality, then we are forever lost. The Buddha said, in essence, "don't take my word for it—look for yourself"[2]. When you awaken you will clearly see that you were in the Garden all along. But for most of us, we will continue dreaming the dream that our ego dreams. The self is parasitized by the ego, which takes command of the sixth sense and so claims both mind and body. This is allowed to happen because we have literally forgotten ourselves. *Time Sutra* is meant as a thread that will help us remember that critical thing we have forgotten. As for me, I sit by the path leading into the Garden, still in the dream, but knowing it is a dream,[3] which is also a way of knowing reality. This isn't enlightenment, but it is some serious brush-clearing that exposes the path, which has been there all along.

I do think that any successful Western Buddhism will look something like *Time Sutra*. Buddhism is a deconstruction of the host culture—it is a tearing down of the framework constructed out of cultural beliefs about the nature of the world. This is what Buddhism does—it removes the distorting screen through which we commonly see the world. The job of Western Buddhism is what *Time Sutra* clumsily attempts. If Buddhism is to have meaning, it will deconstruct the modern paradigm. But this requires an intimate knowledge of the failings of logic and reason in the timespace continuum.

Buddhism accepts science—the Dalai Lama embraces science. Science is the Western paradigm. And many Western scientists embrace Buddhism because it is reasonable and empirical. But Westerners who also embrace their scientific ideology don't recognize that all their unnamed presumptions will be unearthed and discarded by the deconstructive nature of Buddhism.

I'm partial to the idea of the Perennial Philosophy, which is accessible in all ages and all cultures,[4] and is always available now. It is a philosophy that is arrived at by anyone who takes up the thread of reason and logically couples reason with sensory experience. The existence of the Perennial Philosophy has long been recognized but never well defined. *Time Sutra* is a modern recitation of this Philosophy, couched in the vernacular of early 21st century science and modern philosophy. With this exposition of *Time Sutra* in mind, go and read the Buddhist or Vedic texts, and keep *Time Sutra* in mind when you read Lao Tzu, Longchenpa, Shankara, or Parmenides.[5] There is a Perennial Philosophy inherent in reality, and this makes it possible for anyone at any time to quiet the mind and begin to remember the self as it was, before it was commandeered by the ego.

I have been chasing this one idea for over twenty-nine years. The single original thing I bring to this discussion is pointing out that the belief, *memory is past,* is the one, rudimentary element that creates not only time, but is also foundational to the entire timespace continuum that circumscribes the way we think about the world. This is the ground that our culture is built on. Additionally, *memory is past* is responsible for the construction of ego, thus the original experience of Being has been appropriated by ego. The self has been stolen, and the soul corrupted.

Time Sutra may not engender enlightenment, but it can elicit the state of lucid dreaming—we can know that we are asleep in the Garden. *Time*

Sutra is a meditation, and through this meditation, we can practice lucidity—it is no esoteric mystery, it is nothing but the application of reason, logic, and observation. We must all acquire it in our own way, so that the understanding becomes our own personal knowledge.

I have a meditation that allowed (or perhaps caused) me to write *Time Sutra*, and through this meditation I still glimpse the beginning—that point where time begins and ends. "The beginning cannot be preserved as beginning; it can only be remembered or forgotten."[6] And I know that the cause of 'beginning and end' is nothing other than thinking that *memory is past*. All we have to do is look, and by looking, know with certainty, that the phenomenon of memory is in the present. Heraclitus and Plato were both saying "look within" and "turn the gaze around". If you sit down and seriously think about it, this is where you end up. Many cultures have realized that they have fallen into time, and because time is a lie, they have developed elaborate rituals and practices to temporarily regain their original timeless nature. Modernity has been lacking a salvation from time, because Modernity was never completed. Montaigne was right, our knowledge begins and ends in *sense*. Decode the true nature of the *sense* of memory and you will have acquired the insight that *Time Sutra* reveals.[7]

Time Sutra solves many of the vital problems of philosophy: it identifies memory as the sixth sense, fills the 'explanatory gap', details the 'nature of time', explains the nature of ego and how it is formed, thus resolving the mind-body problem, and finally, brings Modernity to a positive conclusion. And, if you accept the conclusion, then *Time Sutra* points out the path to enlightenment.

The last two chapters of *Time Sutra*, particularly the last, devolve into metaphor, because all that is left to be said lies somewhere beyond language. The reader who has understood what has been said, and knows the arguments and agrees with their conclusions, must feel some compulsion to turn the gaze fully around, and look at the reality. This is where the path steepens (or perhaps, a better metaphor), this is where the

199

path is so overgrown, because the path has not been used since the beginning of time. The source of the path originates from the mental construction that we think of as *time*, but the beginning is always timelessly here.

Traditional meditation is effective in gaining emotional detachment from your thoughts, and is a way of diminishing ego. Because, if you don't take your thoughts seriously enough to become involved in them (and most common meditative practices are learning to dispassionately observe the tumultuous thoughts that course through our mind), the result is that ego no longer dominates. Meditation works when it shuts off the mental control from ego, because it is ego that is so passionately involved in thinking. The program running on the mind is the ego-construct, and meditation interrupts that program. The solution is not to try to ignore *thinking,* or shut it off, but to discover why our thinking is so wrong. Sitting meditation works, but it doesn't, in itself, get to the crux of the problem—we need to know why we think this way—then, when we stand up from the meditation mat there is not a lapse back into egoic mind.[8]

If *Time Sutra* is to be a mnemonic tool, it must be held in memory.[9] However, it really only comes down to exchanging the belief that *memory is past* for the obvious fact that *memory is present*. But memorization does no good if you cannot grasp the argument. Understanding is the capacity to experience memory timelessly. This is the same as running an egoless program that interprets all of memory to be present experience, which in turn, has the profound effect of timeless experience—experienced by no *one*. The phenomena are no longer divided into what is experienced and the one who has the experience. The timeless, ego-less experience is by its very nature, an experience of unity. We are entangled in the way we think, but reason can untangle what unreason has fallen into.

I know that it was very presumptuous of me to have written this book. I am standing against the current modern era, but I also maintain that the modern era has an enormously important truth to teach us. This truth has not yet been learned, and as long as it goes unlearned we are stuck in modernity. The science that modernity has granted us makes it explicit that so little of what modern humans do is sustainable. Humans often understand this but don't comprehend how it could possibly be any different.

The change that might be best is to stop the generation of the new for the sake of more new things, and go back and sample the old, and take the very best of the present—we might be able to make the garden sustainable for humans. The first step is to think our way back to the beginning. Once there, there is nothing further that need be done. The problem all along with modernity was the doing, not the non-doing. It has been pointed out that we don't need action, we need theory.[10]

Time Sutra provides the explanation of time. Time is not innate, it is a cultural distortion of memory. The distortion was first recognized over two and a half millennia ago and humans have continued to ponder the problem ever since. Modern thought, regarding this distortion, is crucial to unraveling the deception in a way that it can be understood by the modern mind. This is the culmination of modernity—this is the natural conclusion that modernity comes to—what else could it possibly be?

Perhaps what has been said will be unacceptable, but not because it is wrong. The logic is unavoidable—we know that we cannot know—we can make assumptions and then claim to know, but that is not the same thing as wisdom.

NOTES

PREFACE

1. Ludwig Wittgenstein, *Philsophical Investigations*, 1958, #309. The fly is free to leave the fly bottle but it is imprisoned by its own habit.

2. For years I wondered how T. S. Eliot had gained such a powerful insight into time. One day, searching through his other writings, I came across the fact that his Harvard doctoral dissertation had been published the year before his death and it was on the philosophy of F. H. Bradley who had written extensively about the philosophy of time. Suddenly, I knew how Eliot had gained such a deep understanding of time. By the way, it is a very well-written explanation of F. H. Bradley's philosophy, but this should be no surprise, after all it was written by T. S. Eliot. T. S. Eliot's dissertation title is *Knowledge and Experience in the Philosophy of F. H. Bradley*. Bradley's first major publication, *Appearance and Reality* was written in 1893 and reveals his insight into time.

3. T.S.Eliot, 1943, *Four Quartets*, p.19.

4. It is difficult, perhaps impossible, to hold the entire time-memory-ego nexus in mind all at once. *Time Sutra* is just a reminder. A mnemonic nudge toward reality. The length of the text of *Time Sutra* is intentionally kept brief because the original intent of the Sutras of Buddhism was as a mnemonic device to help the student find and *remember* their way through a very complex problem, the problem of ascertaining reality. This is a way of 'calling to memory' an understanding. As it is with telling any true story, there is a lot I have left out, but I hope what is said is sufficient to get the gist of what *Time Sutra* is trying to communicate.

5. David R. Griffin, 1986, *Physics and the Ultimate Significance of Time*, back cover. "The dominant 20th century view, supported by Einstein and many of the founders of quantum theory, implies that time is ultimately unreal."

6. The *Past Hypothesis* is an assumption that must necessarily accompany the Big Bang Theory of the cosmos. This concept will be explained in chapter 4.

7. The *Representative Theory of Memory* is another type of 'past hypothesis', which assumes memory is the evidence of *Past*. All of our language and thinking are infused with this assumption. This concept will be further explained in chapters 2 and 4.

8. The 'disenchantment of the world' is caused by the self-other split that is made so visible by modernity.

    Here are some pertinent comments from others who have reflected on the nature of *disenchantment*: "At the root of modernity and its discontents lies what Max Weber called 'the disenchantment of the world'." (David R. Griffin, 1988, *The Reenchantment of Science: Postmodern Proposals*, p. 1).

    "Whereas this disenchantment of nature was originally carried out (by Galileo, Descartes, Boyle, and Newton and company) ..." (David R. Griffin, 1988, *The Reenchantment of Science: Postmodern Proposals*, p. 3).

    "The program of the Enlightenment was the disenchantment of the world; the dissolution of myths and the substitution of knowledge for fancy." (Max Horkheimer and Theodor W. Adorno, 2002, *Dialectic of Enlightenment*, p. 3).

    "The disenchantment of the world is the extirpation of animism." (Max Horkheimer and Theodor W. Adorno, 2002, *Dialectic of Enlightenment*, p. 5).

    "Thought and feeling—the psychological—are now confined to minds. This follows our disengagement from the world, its 'disenchantment', in Weber's phrase." (Charles Taylor, 1989, *Sources of the Self*, p. 186).

"For Buddhism, the only solution to this situation is a spiritual one: to deconstruct the elusive sense of a duality between such an alienated self and its objectified, disenchanted world. To conclude we need to see how Buddhism relates this deconstruction of self to the deconstruction of causality, since the problem of means and ends depends upon our more basic notion of cause and effect." (David Loy, 2002, *A Buddhist History of the West*, p. 191).

PART I: BEGINNINGS

Chapter One: Introduction to Time

1. T. M. P. Mahadevan in H. S. Prasad, ed., 1992, *Time in Indian Philosophy*, p. 546.
2. W.K. Mundle in Paul Edwards, ed. 1972, *The Encyclopedia Of Philosophy*, Vol. 8, p. 138.
3. Jean-François Lyotard, 1991, *The Inhuman*, p. 70.
4. Lewis Mumford, 1934, *Technics and Civilization*, p. 14.
5. David Wood, 1989, *The Deconstruction of Time*, p. xxxv.
6. Samuel Alexander in Charles M. Sherover, ed., 1989, *The Human Experience of Time*, p. 239.
7. Jean François Lyotard, 1991, *The Inhuman*, p. 107.
8. Maurice Nicoll, 1953, *Living Time, p. 66.*
9. British astronomer Arthur Eddington coined the term *arrow of time* in 1927 to describe the idea of the one-way direction of time.
10. Entropy is the tendency of any ordered system to become increasingly disordered over time. See also *entropy* in the Glossary and the discussion of entropy in Chapter 4.
11. Huw Price, in *Nature,* 348, Nov. 22, 1990, p. 356.

12. Julian Barbour, 1999, *The End Of Time*, p. 308.

13. Brian Greene, 2004, *The Fabric of the Cosmos*, p. 6.

14. Peter Coveney and Roger Highfield, 1990, *The Arrow of Time*, p. 297.

15. Henry Mehlberg, 1980, *Time, Causality, and the Quantum Theory*, vol. 2, pp. 200-1.

16. Larry Dossey, 1982, *Time, Space, and Medicine* p. 41.

17. Robert E. Ornstein, 1997, *On The Experience Of Time*, p. 17.

18. Christoph Hoerl and Teresa McCormack, eds. 2001, *Time and Memory*, p. vii.

19. J. Krishnamurti, 1969, *Freedom From The Known*, p. 44.

20. Luis Jorges Borges, 1998, 'Book of Sand', *Collected Fictions*, p. 441.

21. Martin Heidegger, 1962, *Being and Time*, p. 479. Heidegger wrote *Being and Time*, his first publication (1927), as a young man. As an old man, he was still searching for what had been covered up in time.

22. Ruth Reyna, quoting Alfred North Whitehead in H. S. Prasad, ed., 1992, *Time in Indian Philosophy*, p. 724.

23. John J. McDermott in Samuel L. Macey, ed., 1994, *Encyclopedia of Time*, p. 482.

24. Franz Brentano, 1988, *Philosophical Investigations in Space, Time and the Continuum*, p. 49.

25. Alfred North Whitehead in Charles M. Sherover, 1989, *The Human Experience of Time*, p. 341.

26. Brian Greene, 2004, *The Fabric of the Cosmos*, p. 127.

27. Jacob Needleman, 2003, *Time And The Soul*, p. 25.

28. Ibid., p. 108.

29. Steven Heine, 1985, *Existential and Ontological Dimensions of Time in Heidegger and Dogen*, p. 147.

30. Lawrence Sklar, quoting Arthur Eddington, in Le Poidevin and MacBeath, eds., 1993, *The Philosophy of Time*, p. 116.

31. Thomas Cleary, trans., 1978, *Original Face: An Anthology of Rinzai Zen*, p. 60.

32. Roger Scruton, 1999, commenting on Spinoza's philosophy in, *Spinoza*, p. 29.

33. Mircea Eliade in H. S. Prasad, ed., 1992, *Time in Indian Philosophy*, p. 110.

34. T. M. P. Mahadevan, quoting Samuel Alexander in H. S. Prasad, ed., 1992, *Time in Indian Philosophy*, p. 596.

35. T. S. Eliot, 1943, 'The Dry Salvages', *Four Quartets*, p. 44.

36. Samuel Beckett, 1965, *Proust: and Three Dialogues with George Duthuit*, p. 11.

37. Ibid., p. 18.

38. Klaus Klostermaier in Samuel L. Macey, ed., 1994, *Encyclopedia of Time*, p. 266.

39. Raimundo Panikkar, quoting Bhartrhari, in H. S. Prasad, ed., 1992, *Time in Indian Philosophy*, p. 21.

40. Raimundo Panikkar,1978, in J. T. Fraser, et al, eds., *The Study of Time III*, p. 702.

41. Jacob Needleman, 2003, *Time And The Soul*, p. 26.

42. Charles Genoud, 2006, *Gesture of Awareness*, p. 148.

43. Aldous Huxley, 1945, quoting Meister Eckhart, *The Perennial Philosophy*, p. 189.

44. Dainin Katagiri, 2007, *Each Moment Is The Universe: Zen and the Way of Being Time*, p. 8.

45. Ibid., p. 38.

46. John Blofeld, trans., 1958, *The Zen Teachings of Huang Po*, p. 58.

47. Nagarjuna and Kaysang Gyatso, The Seventh Dalai Lama, Jeffrey Hopkins, et al., trans., 1975, *The Precious Garland and The Song of the Four Mindfulnesses*, p. 34, verses 113-114.

48. Somerset Maugham, 1984, *The Razor's Edge*, p. 269.

207

49. Bertrand Russell (commenting on Hegel's philosophy), 1945, *A History of Western Philosophy*, p. 735.

50. Joan Stambaugh, 1987, *The Problem of Time in Nietzsche*, p. 59.

51. Eckhart Tolle, 1999, *The Power of Now*, p. 49.

52. H. S. Prasad, ed., 1991, *Essays on Time in Buddhism*, p. xxiii.

53. David Wood, 1989, *The Deconstruction of Time*, p. 1.

54. Elizabeth Grosz, 1999, quoting Derrida, *Becomings: Explorations in Time, Memories, and Futures*, p. 1.

55. Craig Callender, 2010, 'Is Time An Illusion?,' *Scientific American*, June 2010, p. 41.

56. Stefan Klein, 2006, quoting Albert Einstein, *The Secret Pulse of Time*, p. 257.

57. Robert E. Ornstein, 1997. *On The Experience Of Time*, p. 16. Ornstein offers this great quote from H. Nichols, taken from, 'The Psychology of Time,' (1891) *Amer. J. Psychol.*, vol. 3, pp. 453-529.

58. Samuel L. Macey, 1991, *Time: A Bibliographic Guide*, p. xviii.

Chapter Two: The Fall

1. St. Augustine, F. J. Sheed, trans., 1942, *Augustine: Confessions*, Book XI, p. 217.

2. The biblical fall is man being thrown out of the Garden by God for eating the 'fruit of knowledge'. The Greek fall from grace is due to *techné* (technology)—this is the myth of Prometheus giving fire to humans. Both are admonitions against the presumption of knowledge.

3. Jean-Jacques Rousseau (1712–1778) was an early modern writer who identified *the* central philosophical problem as a 'fall into time'.

4. Along with the Sacred Time and Great Time, another synonym is the term Dreamtime, used by the Australian Aboriginals. All of these terms are used in contrast to profane time or mundane time.

5. Max Oelschlaeger, 1991, quoting Loren Eisley, in *The Idea Of Wilderness*, p. 333.

6. Yehuda Thomas Raddy and Haim Shore, 1985, *Genesis: An Authorship Study in Computer-Assisted Statistical Linguistics.*

7. Richard Geldard, 2000, *Remembering Heraclitus*, p. ix.

8. Karen Armstrong, 2006, *The Great Transformation*, pp. 260-1.

9. Ibid., p. xii.

10. Martin Heidegger, 1998, *Parmenides*, p. 24.
    Heidegger presented a 1942–3 lecture series at Freiberg University, published posthumously in 1982 as vol. 54 of his collected works. English translation, *Parmenides,* 1992, Indiana University Press.

11. "In Parmenides' rejection of the 'way of the mortals', it was seen that sense-experience in itself did not seem to be blamed for their mistakes. It was mortals' habitual misinterpretation of sense experience which caused them to fall into self-contradiction." (C. C. W. Taylor, ed., 1997, *Routledge History of Philosophy*, vol. 1, p. 145).

12. *Aletheia* at the time of Parmenides writing was a word that meant *Truth*, but was also the name for the Goddess of Memory, and when translated literally, the word means *not-forgetting*.

13. Samuel L. Macey, 1994, *Encyclopedia of Time*, p. 536.

14. A 19th century contraption to catch and drown flies based on the fly's instinctive nature to fly upwards and toward the light.

15. Jorges Luis Borges, 1998, 'Shakespeare's Memory' in *Collected Fictions*, p. 506.

16. Samuel L. Macey in Paul Edwards, ed., 1972, *The Encyclopedia of Philosophy*, vol. 5, p. 272.

17. Knowledge of past "is based ultimately on memory" is a conclusion reached by B. Russell, C. D. Broad and H. H. Price in Paul Edwards, ed., 1972, *The Encyclopedia Of Philosophy*, vol. 5, p. 272, which reduces knowledge of past to the assumption that memory does indeed represent past.

18. "Critics of the representative theory have generally followed Reid in charging that this theory, far from explaining how we can have knowledge of the past, has the consequence that such knowledge is impossible." (Paul Edwards, ed., 1972, *The Encyclopedia Of Philosophy*, vol. 5, p. 269).

Chapter Three: Einstein and Piaget

1. T. S. Eliot, 1943, 'East Coker', *Four Quartets*, p. 23

2. Piaget credits Einstein for the direction of his research on childhood psychological development. Piaget begins the foreword of *The Child's Conception of Time* with the statement, "This work was promoted by a number of questions kindly suggested by Albert Einstein more than fifteen years ago when he presided over the first international course of lectures on philosophy and psychology at Davos." (Jean Piaget, 1969, *The Child's Conception of Time*, p. vii). I have seen the date of the Davos meeting reported to be 1928, 1930, and 1931, but Piaget does say that it was the "first" of the Davos meetings.

3. At age twenty-five Piaget became director of the Rousseau Institute in Geneva where he pursued his work in experimental psychology.

4. Piaget characterizes Einstein's questions as follows: "Is our intuitive grasp of time primitive or derived? Is it identical with our intuitive grasp of velocity? What if any bearing do these questions have on the genesis and development of the child's conception of time?" (Jean Piaget, 1969, *The Child's Conception of Time*, p. vii). For the purposes of *Time Sutra*, the first question proves to be central.

5. *The Child's Conception of Time* is the title of the 1969 English translation (reprinted 1971) of the original 1946 French publication.

6. Jean Piaget, 1971, *The Child's Conception of Time*, p. vii.

7. "Piaget was criticized by many later researchers for drawing his conclusions more from conversations than from experiments, but no one denies that the idea of time is far from intrinsic. It takes quite a bit of effort to learn." (Stefan Klein, 2006, *The Secret Pulse of Time*, pp. 138-9).

8. Samuel L. Macey, ed., 1994, *Encyclopedia of Time*, p. 466.

9. William J. Friedman, 1990, *About Time: Inventing the Fourth Dimension*.

10. Ibid. p. 43

11. Ibid. p. 43.

12. " ... it is by learning to tell stories to others that a child learns to tell stories to himself and thus to organize his active memory." (Jean Piaget, 1971, *A Child's Conception of Time*, p. 272).

13. Freud coined the term "childhood amnesia," but the phenomenon was recognized as early as 1893 by psychologist Caroline Miles.

14. "In making memory the animating property of the self, Diderot was following Locke and Condillac, but he did so in his own way." (Jerrold Seigel, 2005, *The Idea of the Self*, p. 191).

15. Ibid. p. 191.

16. Daniel L. Schacter, 1996, quoting St Augustine, *Searching for Memory*, p. vii.

17. Shaun Gallagher and Jonathon Shear, 1999, *Models of the Self*, p. x.

18. "Locke takes up the question of personal identity. What you mean by it is just your chain of particular memories." (William James, 1987, *Varieties of Religious Experience: Writings, 1902–10*, p. 398). This is very similar to Edward S. Casey's "we are what we remember".

19. Daniel M. Wegner, 2002, citing Grice 1941; Perry 1975, *The Illusion Of Conscious Will*, p. 264.

20. "One phenomenon that has been observed in several different experimental tasks is that children of 3 years or younger seem to be unaware of former states of the self." (Katherine Nelson in Daniel L. Schacter and Elaine Scarry, 2001, *Memory, Brain, and Belief*, p. 273).

21. "How can we 'see' through our past life if it consists in a mere succession of first order mental states and episodes, neither unified nor bound together by any reflection? In this perspective, episodic memory requires a present exercise of self consciousness, but it also requires that the capacity of self consciousness be in place from the remembered experience to the present. Since such a capacity appears to emerge at around the age of 3 or 4, the phenomena of childhood amnesia would be explained by the presence of a 'block', at about that time, beyond which episodic memory is simply blind." (Jerome Dokic, in Hoerl and McCormack, eds., 2001, *Time And Memory: Issues in Philosophy and Psychology*, p. 231).

22. Ulric Neisser and Ira E. Hyman, 2000, *Memory Observed*, p. 298.

23. "By age four years children enjoy and actively participate in talking with others about their lives. Sharing memories provides an important context for social participation as well as self reflection." (Katherine Nelson, in Daniel L. Schacter and Elaine Scarry, 2001, *Memory, Brain, and Belief*, p. 267).

24. Richard Noll and Carol Turkington, *1994, Encyclopedia of Memory and Memory Disorders*, p. 21.

25. "For us believing physicists, the division into the past, present and future has merely the meaning of an albeit obstinate illusion." (H. Deiter Zeh, 1999, quoting Einstein, *The Physical Basis of the Direction of Time*, p. 195).

26. "Einstein saw at once that if Gödel was right, he had not merely domesticated time: he had killed it. Time, 'that mysterious and self-contradictory being,' as Gödel put it, 'which, on the other hand, seems to form the basis of the world's and our own existence,' turned out in the end to be the world's greatest illusion. In a word, if Einstein's relativity was real,

212

time itself was merely ideal." (Palle Yourgrau, 2005, *A World Without Time*, p. 7).

27. The exact quote: "*Time*, he [Gödel] told Wang years later, remains, even after Einstein, *the* philosophical question." (Palle Yourgrau, *A World Without Time*, 111).

Chapter Four: Entropy, Time's Arrow, and The Past Hypothesis

1. "As the philosopher Alcmaeon poignantly but obscurely puts the point in his sole surviving fragment." (Samuel L. Macey, ed., 1994, *Encyclopedia of Time*, p. 494). Note: Others have attributed additional fragments to him.

2. The term 'flow' used to characterize the 'passing' of time is at least as old as Heraclitus' allusion to change as being a river into which "one cannot step twice". More recently 'flow' is sometime used to mean nearly the opposite—as a sense of the timeless associated with the involvement and enthralment with present being. "One of the most common descriptions of optimal experience is that time no longer seems to pass the way it ordinarily does." (Mihaly Csikszentmihalyi, 1990, *Flow: The Psychology of Optimal Experience*, p. 66). However, *Time Sutra* uses the term 'flow' in the same way Heraclitus meant it.

3. Many philosophers in the latter half of the 20th century attempted to devise a tenseless theory of language, but the result is a near-impenetrable quagmire of words.

4. Jorges Luis Borges, 1986, 'A New Refutation of Time'*, in *Labyrinths: Selected Stories and Other Writing*, p. 218.

5. John McTaggart, 1927, *The Nature of Existence*, vol. 2.

6. The details of McTaggart's argument lie outside our story. *The Cambridge Dictionary of Philosophy* (p. 527) summarizes his argument: "In the most celebrated part of his philosophy, he argued that time is unreal by claiming that time presupposes a series of positions, each having the incompatible qualities of past, present, and future. He thought that attempts to remove the incompatibility generate a vicious infinite regress."

7. The term 'Arrow of Time' was first used by Arthur Eddington in 1927.

8. "Laplace (1749–1827) argued that given only a knowledge of the total mechanical state of the Universe at any moment of time 'nothing would be uncertain and the future, as the past, would be present to (our) eyes'." (W. F. Bynum, et al., eds., 1981, *Dictionary of the History of Science*, p. 96).

   "All the successful equations of physics are symmetrical in time. They can be used equally well in one direction of time as in the other. The future and the past seem physically to be on a completely equal footing. Newton's laws, Hamilton's equations, Maxwell's equations, Einstein's general relativity, Dirac's equation, the Schrödinger's equation,—all remain effectively unaltered if we reverse the direction of time." (Roger Penrose, 1989, *The Emperor's New Mind*, p. 392).

   " ... you might think that an inspection of the basic micro-physical laws would readily reveal time's arrow. It will not." (Bradley Dowden, *Internet Encyclopedia of Philosophy*, <http://www.iep.utm.edu/time/>).

9. Brian Greene, 2004, *The Fabric of the Cosmos: Space, Time, and the Texture of Reality*, pp. 144-5.

10. Clausius' formulation of the first two laws of thermodynamics:
    First Law: The energy of an isolated system remains constant.
    Second Law: the entropy of an isolated system strives toward a maximum.
    W. F. Bynum, et al., eds., 1981, *The Dictionary of the History of Science*, p. 180.

11. "Only the second law of thermodynamics indicates clearly a direction of time." C. F. V. Weizsacker quoted in J. T. Fraser, ed., 1978, *The Study of Time III*, p. 135.

12. *Time Sutra* will not give a formulation of the Second Law because as David Hull states: "In a recent thermodynamics text, Truesdell (1984) identifies several different 'Second Laws,' and the physicist-philosopher Mario Bunge (1986, p. 306) compiled a list of 'twenty or so ostensible inequivalent but equally vague formulations of 'the' Second Law. The appearance of consensus, which so impresses outsiders, is enhanced by the insistence of those on the inside that one true view does exist—theirs." (David Hull, in B. Weber, D. Depew and J. Smith eds., 1988, *Entropy, Information, and Evolution*, p. 3).

13. Second Law: the entropy of an isolated system strives toward a maximum. W. F. Bynum et al., eds., 1981, *The Dictionary of the History of Science*, p. 180.

    Entropy: A mathematical factor that is the measure of the unavailable energy in a thermodynamic system (*Merriam-Webster Collegiate Dictionary*).

14. Sean Carroll, 2010, *From Eternity to Here*, p. 43.

15. "Fundamentally, then, the universe as a whole could very well be time-symmetric, running all the way from minus eternity to plus eternity with no preferred direction, no arrow of time. Indeed, the whole notion of beginning is meaningless in a time-symmetric universe." (Victor J. Stenger, 2000, *Timeless Reality*, p. 332).

16. *Future* and *history* are contained in quotes because they are theory-laden terms that have nothing to do with timelessness. Their meaning is derived from thinking *memory is past*.

17. H. Bondi, quoted in J. T. Fraser et al, eds., 1978, *The Study of Time III*, p. 136.

Others have also pointed out this same logical problem: "This is an enormous paradox. How can the result of many elementary processes, each perfectly reversible, be in a world in which essentially nothing is reversible." (David Park, 1980, *The Image Of Eternity: Roots of Time in the Physical World*, p. 51).

"For the main argument against the derivability of the entropy law is that (as Poincaré remarked) irreversible conclusions cannot be derived from a reversible theory such as statistical mechanics." (Karl Popper, 2002, *The World of Parmenides*, p. 186).

18. David A. Park, in Samuel L. Macey, ed., 1994, *The Encyclopedia of Time*, p. 44.

Here's the crux of the problem with entropy. People generally don't understand that entropy calculations exclude history, and therefore have no basis for predictions about the future. People only look at the mathematics, and indeed, if you accept that the Second Law is predicting the future, then the overwhelming probability *is* that entropy will be higher in the future.

Here is the short-form of the argument (adapted from *The Encyclopedia of Time*, p.44) that exposes why the "illusion of large numbers" is so convincing and at the same time so wrong: Imagine a one-meter sealed tube into which a specified number of gas molecules are injected instantaneously into the right half of the tube. The right end of the tube is then 'measured' every 100th of a second for the presence of molecules. This mathematical thought experiment then asks: how long will it take for a particular number of molecules to again simultaneously end up in the right half of the tube after injection. Beginning with just *one* molecule released in the right half of the tube—it can be predicted statistically that on average this same condition will be detected every 1/50th of a second (i.e. every 2nd measurement). With ten gas molecules the initial condition will be measured on average every ten seconds. With a total of twenty particles, they will recur on the right side every three hours. Fifty molecules will on average reassemble on the right side of the tube every 350,000 years. And one hundred randomly moving molecules will again arrange themselves together into the right side of the tube only once in tens of billions of years. Of course in the real world we are not dealing with 100 molecules—a single liter of gas at room temperature and pressure contains Avogadro's number of molecules, which is a whole number with 23 digits. Thus, for an aggregation of randomly moving gas molecules, the mathematics makes it look certain that the future will be of higher entropy. Except, this theory cannot predict anything because it knows nothing about the past of the system.

The telling point is that the Second Law predicts both the past and the future equally well, except both predictions are *identical*—the past and the future are indistinguishable. If you don't try to force a biased interpretation onto the Second Law, it is time symmetric.

19. Again, this is Clausius' formulation of the Second Law. (W. F. Bynum, et al., eds., 1981, *The Dictionary of the History of Science*, p. 180).

20. "Thus, *not only is there an overwhelming probability that the entropy of a physical system will be higher in what we call the future, but there is the same overwhelming probability that it is higher in what we call the past.*" (Brian Greene, 2004, *The Fabric of the Cosmos*, p. 160).

21. Ibid., p. 161

22. "The most mysterious thing about time is that it has a direction: the past is different from the future. That's the *arrow of time* ... " (Sean Carroll, 2010, *From Eternity to Here*, p. 2).

23. Ibid., p. 2.

24. "Remarkably, a single concept underlies our understanding of irreversible processes: something called *entropy*, which measures the 'disorderliness' of an object or conglomeration of objects. Entropy has a stubborn tendency to increase, or at least stay constant, as time passes—that's the famous Second

217

Law of Thermodynamics. And the reason why entropy wants to increase is deceptively simple: there are more ways to be disorderly than to be orderly, so (all else being equal) an orderly arrangement will naturally tend toward increasing disorder." (Sean Carroll, 2010, *From Eternity to Here*, p. 2).

25. "Within our observable universe, the consistent increase of entropy and the corresponding arrow of time cannot be derived from the underlying reversible laws of physics alone. They require a boundary condition at the beginning of time. To understand why the Second Law works in our real world, it is not sufficient to simply apply statistical reasoning to the underlying laws of physics; we must also assume that the observable universe began in a state of very low entropy. David Albert has helpfully given this assumption a simple name: the *Past Hypothesis*." (Sean Carroll, 2010, *From Eternity to Here*, p. 176).

26. "Indeed, the Past Hypothesis is more than just "allowed"; it's completely necessary, if we hope to tell a sensible story about the universe. Imagine that we simply refused to invoke such an idea and stuck solely to the data given to us by our current macrostate, including the state of our brains and our photographs and our history books. We would then predict with strong probability that the past as well as the future was a higher entropy state, and that all of the low entropy features of our present condition arose as random fluctuations. That sounds bad enough, but the reality is worse. Under such circumstances, among the things that randomly fluctuate into existence are all pieces of information we traditionally use to justify our understanding of the laws of physics, or for that matter all of the mental states (or written down arguments) we traditionally use to justify mathematics and logic and the scientific method. Such assumptions, in other words, give us absolutely no reason to believe we have justified anything, including those assumptions themselves." (Sean Carroll, 2010, *From Eternity to Here*, pp. 183-4).

27 "I feel that the possibility of time-reversal has been widely neglected for the wrong reason—a deep prejudice that time can only pass from past to future. Evidence for this cannot be found in physics. The only justification for a belief in directed time is human experience, and human experience once said the world was flat." (Victor Stenger, 2000, *Timeless Reality*, p. 12).

"My thesis is not that time reversibility is required to understand the universe, rather, as happened with the Copernican solar system, once we get over our anthropocentric prejudices time reversibility provides us with a simpler and more economical picture of that universe. And it is on that basis, not proof, that we can rationally conclude that time is reversible in reality." (Victor Stenger, 2000, *Timeless Reality*, p. 208).

28. "This now tells us how precise the Creator's aim must have been: namely an accuracy to one part in 10 to the 10th power to the 123rd power." (Roger Penrose, 1989, *The Emperor's New Mind*, p. 445). Penrose goes on to say that there are more zeros in this number than elementary particles in the universe.

"Drawing on the fact that entropy is related to probability, he [Roger Penrose] arrives at a figure of 1 in [10 to the 10th to the 123rd power]—in other words, at the conclusion that in purely statistical terms, the actual early universe was special, or unnatural, to this stupendous degree." (Huw Price, 1996, *Time's Arrow and Archimedes' Point*, p. 83).

Based on thermodynamics, the following is the calculated state of current entropy: "The probability for the present, almost homogeneous universe of 2.7K is therefore a mere 10 to the minus 123rd power." (H. Deiter Zeh, 1999, *The Physical Basis for the Direction of Time*, p. 153).

29. *Memory-is-past* is an assumption very closely related to the *Past Hypothesis*. The first assumes that 'memory is past', while the second assumes past, but also unknowingly assumes that memory represents 'before now', thus assuming time.

The psychological fallacy includes the 'causal arrow' and the 'perceptual/psychological arrow'. The statistical fallacy includes the 'thermodynamic', 'cosmological', 'radiative', 'particle/weak', and 'quantum' arrow of time.

30. Sean Carroll, 2010, quoting Stephen Toulmin, *From Eternity to Here*, p. 315.

Chapter Five: Parmenides

1. Mircea Eliade, 1963, *Myth and Reality*, pp. 87-8.

2. The full quote is as follows: "Still, the overall outline of the poem is clear; the sequence of the fragments is (with relatively few uncertainties) settled; and the main fragment (B8) constitutes the longest, most continuous, and most cohesive text from the period before the Sophists." (Alexander P. D. Mourelatos, 1970, *The Route of Parmenides*, p. xii).

3. "Of all Presocratic philosophers it was Parmenides who exerted the deepest influence on Plato's thought." (Charles H. Kahn, 2009, *Essays on Being*, p. 195).

4. *Truth* is the common translation of the Greek word *Aletheia*, but the term has different connotations. The literal meaning comes from the *'A'* of Aletheia meaning *not,* and *'lethe'* meaning *forgetting*, thus *Aletheia* literally means *not-forgetting*. In Parmenides' time, the term was also used as a personification of the Goddess of Memory.

5. *Belief* is a translation of the Greek word *Doxa*, also translated as *Opinion* or *Seeming*.

6. The Way of Belief/Opinion, as presented in *On Nature*, was probably the existing cosmology and claimed knowledge of Parmenides' time. He clearly stated that it is wrongheaded thinking, so it plays no role in understanding Truth. There has been much debate about why it was included in the first

place. Aletheia stated clearly that she will fully inform Parmenides of the beliefs of the wrongheaded in *Doxa* (Way of Belief/Opinion). Parmenides is simply exposing the errors of oppositional and dualistic thinking.

"Although much of this part [The Way of Opinion] has been lost, the doxography suggests that it presented a cosmology along the lines of Parmenides' predecessors, constructing the world out of opposites which explain change. Clearly, Parmenides thought that all of this was false — or, worse, unintelligible — and it is decidedly odd that he would write it at all." (Phillip Turetzky, 1998, *Time*, p. 11).

"It seems to be, not sense perception itself which is at fault here, but people's lazy habits in selecting and interpreting the information given by sense perception ... It is reason that must dictate how sense perception is to be understood, and not the other way around." (C. C. W. Taylor, ed., 1997, *Routledge History of Philosophy*, vol. 1, p. 138).

7. Paul Edwards, ed., 1972, *The Encyclopedia Of Philosophy*, vol. 6, p. 49.

8. "The chariot ride of the narrator in the introduction has usually been taken as an allegory of Parmenides' own intellectual odyssey, and of the framework with which he starts." C. C. W. Taylor, ed., 1997, *Routledge History of Philosophy*, vol. 1, p. 150).

9. David Gallop, *Parmenides of Elea,* p. 49, cites three translations: "leaving the House of Night for the light"; "hasting to convey me into the light"; "hastened to escort me towards the light."

10. David Gallop translates: "coaxing her with gentle words". (Ibid., p. 51).

11. The goddess who narrates the poem is unnamed in the surviving text but it seems obvious that she is *Aletheia*, as Heidegger said, "The goddess is the goddess 'truth.' 'The truth'—itself—is the goddess." (Martin Heidegger, *Parmenides*, p. 5).

Marcel Detienne also implies that the goddess is Aletheia when he comments, "In the sixth century B. C., Truth, *Aletheia*, figured as one of the intimate companions of the goddess who greeted Parmenides and guided him to 'the unshakable heart of the perfect circle of truth'." (Marcel Detienne, *1999, Masters of Truth in Archaic Greece*, p. 15).

12. "*Aletheia* is a kind of double to *Mnemosyne*. The equivalence between the two powers is borne out on three counts: equivalent meanings (*Aletheia* and *Mnemosyne* stand for the same thing); equivalent positions (in religious thought *Aletheia*, like *Mnemosyne*, is associated with experiences of incubatory prophecy); and equivalent relationships (both are complementary to *Lethe* [forgetting])." (Marcel Detienne, 1999, *Masters of Truth in Archaic Greece*, p. 65).

"Like *Mnemosyne, Aletheia* is the gift of second sight: an omniscience, like memory, encompassing the past, present, and future." (Marcel Detienne, 1999, *Masters of Truth in Archaic Greece*, p. 65).

13. Martin Heidegger, 1962, *Being and Time*, p. 222.

14. The full quote is as follows: "At the end of the journey we come to the remarkable imagery opening Parmenides' poem and his meditation on Being: 'a chariot journey guided by the Daughters of the Sun, a way reserved for the knowing man, a path that leads to the Gates of Day and Night, a goddess who reveals true knowledge,' and the obligation to choose between the world of *being* and the world of *opinion*." (Foreword by Pierre Vidal-Naquet in Marcel Detienne, 1999, *Masters of Truth in Archaic Greece*, p. 9).

15. Pythagoreanism comes very close to a cult of numerology and memory.

16. Alexander Mourelatos offers an example of this type of thinking: "My suggestion, then, is that Parmenides uses old words, old motifs, old themes, and old images precisely in order to think new thoughts in and through them." (Alexander P. D. Mourelatos, 1970, *The Route of Parmenides*, p. 39)

17. Robert Audi, 1999, *The Cambridge Dictionary of Philosophy*, p. 647.

18. The Prologue is generally considered an allegory, although some have argued that the allegory had not yet been invented and maintain that the Prologue is not an allegory.

    "Parmenides might well claim to be the founder of epistemology: at least he is the first philosopher to make a systematic distinction between knowledge and belief. (Anthony Kenny, *2010, A New History of Western Philosophy,* p. 118).

19. "And in the dialogues [of Plato] Parmenides is the one and only interlocutor who is allowed to defeat Socrates in argument." (Charles H. Kahn, 2009, *Essays on Being*, p. 195).

20. "The development of Western philosophy was once said by A. N. Whitehead to have consisted in a series of footnotes to Plato. In a similar vein, and with hardly more exaggeration, Plato's own writings might be said to have consisted in footnotes to Parmenides of Elea." (David Gallop, trans., 1984, *Parmenides of Elea*, p. 3)

    "Parmenides is not, as some have said, the father of idealism; on the contrary, all materialism depends on his view of reality." (W. K. C. Guthrie, 1965, quoting Burnet, *A History of Greek Philosophy,* vol. II, p. 25).

21. "Indeed, all philosophical and scientific systems that posit principles of conservation (of substance, of matter, of matter-energy) are inalienably the heirs to Parmenides' deduction." (Robert Audi, 1999, *Cambridge Dictionary of Philosophy*, p. 647).

22. The extended quote: "Memory appears to predate any consciousness of the past and any interest in the past as such. At the dawn of Greek civilization, a sort of intoxication before the power of memory is perceptible, but the memory in question has quite a different orientation from our own, and it serves different ends." (Jean-Pierre Vernant, 2006, *Myth and Thought Among the Greeks*, pp. 136-7).

23. Martin Heidegger, 1976, *What Is Called Thinking?*, p. 140.

24. Plato. Edith Hamilton and Huntington Cairns, eds., 1961, *The Collected Dialogues of Plato*, 'Theaetetus' 184a.

    Here is another interpretation of that same passage: "So I'm afraid that we'll fail to understand what he said and that we'll fall even far shorter of what he had in mind when he said it." (Panagiotis Thanassas, quoting Socrates from the Theaetetus, 2007, *Parmenides, Cosmos, and Being—a Philosophical Interpretation*, p. 9).

25. "The Greek word anamnesis means remembering or recollection and is the basis of Plato's theory of knowledge and wisdom." (Richard Geldard, 2000, *Remembering Heraclitus*, p. ix).

26. Richard Velkley characterizing Heidegger. (Richard I. Velkley, 2002, *Being After Rousseau*, p. 149).

27. *Routledge History of Philosophy*, v. 1, prefers translating *aletheia* as *reality*. This works, but the specific connection to the reality of memory is severed. "This conclusion that *aletheia*, in the sense of 'reality', is the intended subject, is central to the interpretation of Parmenides to be presented here." (C. C. W. Taylor, ed., 1997, *Routledge History of Philosophy*, vol. 1, p. 134).

    "But in Parmenides, the translation 'reality' for Parmenides' *aletheia* must be insisted on, in order to bring out the essential point: what is referred to here is not anything (words, speech, thoughts) ..." (C. C. W. Taylor, ed., 1997, *Routledge History of Philosophy*, vol. 1, p. 132).

28. "Parmenides' maxim was that only being—what is—can exist." (Paul Edwards, ed., 1972, *The Encyclopedia Of Philosophy*, vol. 3, p. 46).

    "Parmenides and the Eleatic School criticized the notion of not-being that theories of physical transformations seemed to presuppose." (Robert Audi, ed., 1999, *The Cambridge Dictionary of Philosophy*, p. 734).

29. Anandita Niyogi Balslev, quoting the Bhagavad Gita, 1999, *A Study of Time in Indian Philosophy*, p. 130.

30. G. E. L. Owen, 'Plato and Parmenides on the Timeless Present' in Alexander P. D. Mourelatos, ed., 1974, *The Pre-Socratics*, p. 276.

31. Here is the full quote: "Historians of science or philosophy who are reluctant to attribute to a great thinker like Parmenides a doctrine as severely unempirical as the illusionary character of the world of change (and a doctrine as difficult to accept as the doctrine that consciousness is the only thing in the universe that actually undergoes change) may perhaps be less reluctant when they see that great scientists, such as Boltzmann, Minkowski, Weyl, Schrödinger, Gödel, and above all Einstein, have seen things in a similar way to Parmenides, and have expressed themselves in strangely similar terms." (Karl Popper, 2002, *The World of Parmenides*, p. 172).

32. Erwin Schrödinger, 1954, *Nature and the Greeks*, pp. 26- 7.

33. Here is the entire quote: "He [Parmenides] also anticipated a popular interpretation of the theory of relativity which sees all events and transitions as already prearranged in a four-dimensional continuum, the only change being the (deceptive) journey of consciousness along its world line. Be that as it may, he was the first to propose a conservation law (*Being is*), to draw a boundary line between reality and appearance (and thus create what later thinkers call a 'theory of knowledge') and to give a more satisfactory foundation for continuity than did 19th and 20th century mathematicians who had to invoke intuition. Using Parmenides' argument, Aristotle constructed a theory of space and motion that anticipated some very deep-lying properties of quantum mechanics and evaded the difficulties of the more customary (and less sophisticated) interpretation of a continuum as consisting of indivisible elements. Parmenides' theory clashes with most modern methodological principles but this is no reason to disregard it." (Paul Feyerabend, 1993, *Against Method*, pp. 43-4).

34. The Block Universe will be described in some detail in chapter 13.

35. " ... it is *aletheia* that lies unsaid at the base of what is said by the Greeks." (Charles H. Seibert, trans., [describing Heidegger's concluding remarks] in Martin Heidegger and Eugen Fink, 1993, *Heraclitus Seminar*, p. viii).

    "I make a proposal: the unthought is *aletheia*." (Martin Heidegger and Eugen Fink, 1993, *Heraclitus Seminar*, p. 161).

36. "The field of the essence of aletheia is covered over with debris. But if that were all, then it would be an easy task to clear the debris and once again lay open this field. The difficulty is that it is not merely covered over with debris; there has been built on it an enormous bastion of the essence of truth determined in a manifold sense as 'Roman'." (Martin Heidegger, 1998, *Parmenides*, p. 53).

37. Marcel Detienne has written an historical preface to Parmenides that details how Aletheia changes to Mnemosyne and then, even the meaning of Aletheia is lost. It is written in the foreword that "In some sense, Detienne's aim is to write a prehistory of Parmenides' poem." (Marcel Detienne, 1999, *Masters of Truth in Archaic Greece*, p. 9). Following are several additional quotes from the same source.

    "Through Simonides' thought and work we can see exactly how *Aletheia* came to be devalued ... Simonides was the first to treat poetry as a profession." (Marcel Detienne, 1999, *Masters of Truth in Archaic Greece*, p. 107).

    "Simonides of Ceos, born circa 557–556 B.C." (Marcel Detienne, 1999, *Masters of Truth in Archaic Greece*, p. 107).

    "However, the devaluation of *Aletheia* can only be understood in relationship to a technical innovation representing another fundamental aspect of Simonides' secularization of poetry. An entire tradition attributes to him the invention of the techniques of memory." (Marcel Detienne, 1999, *Masters of Truth in Archaic Greece*, p. 109).

"Through a direct vision or through memory, the poet entered the Beyond, gaining access to the invisible. Memory was the basis of poetic speech as well as the poet's special status. But with Simonides, memory became a secularized technique, a psychological faculty available to all via definite rules that brought it within everyone's grasp." (Marcel Detienne, 1999, *Masters of Truth in Archaic Greece*, p. 110).

"But for Simonides, the new function of memory could not be separated from a new attitude toward time, an attitude diametrically opposed to the religious sects and the philosophicoreligious circles. Whereas the Pythagorean Paron saw time as a power of oblivion, from which the only possible escape was memory, an esthetic and spiritual exercise, for Simonides, time was, on the contrary, 'the best of things,' ... " (Marcel Detienne, 1999, *Masters of Truth in Archaic Greece*, p. 110).

"Here one can also detect an inevitable link between the secularization of memory and the decline of *Aletheia*. Cut off from its source, *Aletheia* abruptly lost value. Simonides rejected it as a symbol of the old kind of poetry. To take its place, he recommended *dokein*, *doxa*.

"This seems to be the first time *Aletheia* was directly opposed to *doxa*; at this point an important conflict arose that would overshadow the entire history of Greek philosophy." (Marcel Detienne, 1999, *Masters of Truth in Archaic Greece*, p. 111).

38. Richard Geldard, 2007, *Parmenides and The Way of Truth*, p. 24.

39. The extended quote: "... *aletheia* is the beginning itself ... The thinker thinks the beginning insofar as he thinks *aletheia*. Such recollection is thinking's single thought. This thought, as the dictum of the thinker, enters into the word and language of the Occident." (Martin Heidegger, 1998, *Parmenides*, p. 163).

"*Aletheia* is *thea,* goddess—but indeed only for the Greeks and even then only for a few of their thinkers. The truth: a goddess for the Greeks in the Greek sense. Indeed." (Martin Heidegger, 1998, *Parmenides*, p. 162).

Chapter Six: Self and Meme

1. Elaine Pagels, 2003, *Beyond Belief: The Secret Gospel of Thomas*, p. 52; The Gospel of Thomas from the Dead Sea Scrolls.

2. Edward S. Casey, 1987, *Remembering*, p. 290.

3. Historically the self/soul has been associated with the heart or breath, but modern science empirically ties the mind/self to the physical brain.

4. Anthony R. McIntosh in H. L. Roediger, Dudai and Fitzpatrick, eds., 2007, *Science of Memory: Concepts*, p. 59.

5. One of the common tests for the recognition of self-awareness is the mirror test. Dogs and cats don't take a self-interest in their mirror reflection. Chimps, orangutans, elephants, dolphins, etc. do see their *self* in a mirror. But only humans understand the self to be in time and space, existing in the interval between birth and death.

6. A more complete quote: "There is reason to think that animals or babies can have phenomenally conscious states without employing any concept of the self. To suppose that phenomenal consciousness requires the concept of the self is to place an implausible intellectual condition on phenomenal consciousness." (Ned Block in Donald M. Borchert, ed., 1996, *The Encyclopedia Of Philosophy*, Supplement, p. 98).

7. The function of the pineal gland was unknown at the time. René Descartes felt that this structure near the center of the brain must contain some important attribute of the self.

8. Cognitive science is the interdisciplinary scientific study that encompasses neuroscience, philosophy, linguistics, psychology, artificial intelligence, anthropology, and education.

9. "If all that matters is the computation, we can ignore the brain's wiring diagram, and its chemistry, and just worry about the 'software' that runs on it." (Daniel Dennett, 2005, *Sweet Dreams: Philosophical Obstacles to a Science of Consciousness*, p. 18).

10. "The concept of cultural replicators—items that are copied over and over—has been given a name by Richard Dawkins (1976), who proposed to call them *memes."* (Daniel Dennett, 2007, *Breaking the Spell*, p. 78). The memes (rhymes with *seems*) act much like genes in that they transfer a unit of information, but rather than information residing in DNA or RNA, the meme resides in memory. The study of memes is called memetics.

11. Semantic memory is general knowledge that is remembered but does not contain any specific attachment to self—there is no historical memory of learning the memory, thus it has no necessary connection with self, other than that it is claimed by self.

    "**semantic memory**: long-term memory for facts, other than autobiographical." (Richard L. Gregory, ed., 2004, *The Oxford Companion to the Mind*, p. xviii).

12. John Kotre, 1996, *White Gloves: How We Create Ourselves Through Memory*, p. 127. The autobiographical memory is composed of those memories that contain time and place, and are the memories that are most closely associated with 'self'.

13. Susan Blackmore in Richard Gregory, ed., 2004, *The Oxford Companion to the Mind*, p. 558.

14. Cultural evolution and cultural drift might be thought of as the consequences of the introduction of new memes into culture. Cultural death might be

thought of as the inability of a culture to propagate the memes that are foundational to that culture.

15. "The computers in which memes live are human brains." (Richard Dawkins, 1976, *The Selfish Gene*, p. 197).

16. "Dawkins (1993) coined the term 'viruses of the mind' to apply to such memeplexes as religions and cults—which spread themselves through vast populations of people by using all kinds of clever copying tricks, and can have disastrous consequences for those infected." (Susan Blackmore,1999, *The Meme Machine*, p. 22).

"Richard Dawkins, who coined the term *meme*, is no friend of religion and has often likened memes—religious memes in particular—to viruses, stressing the capacity of memes to proliferate in spite of their deleterious effects on their human host." (Daniel Dennett, 2007, *Breaking the Spell*, p. 184).

"Memes for blind faith have their own ruthless ways of propagating themselves. This is true of patriotic and political as well as religious blind faith." (Richard Dawkins, 1976 *The Selfish Gene*, p. 198).

17. Susan Blackmore in Richard Gregory, ed., *The Oxford Companion to the Mind*, p. 558.

The full quote: "The vast majority of memes are not viruses but are the very foundation of our lives, including all the arts and sports, transport and communication systems, political and monetary systems, and science."

What Blackmore does not say is that political and monetary systems, along with the sciences, often do turn into ideological convictions, thus becoming mental viruses.

18. "Memetics provides a new way of looking at the self. The self is a vast memeplex—perhaps the most insidious and pervasive memeplex of all ... [it] permeates all our experience and all our thinking so that we are unable to see

it clearly for what it is—a bunch of memes." (Susan Blackmore, 1999, *The Meme Machine*, p. 231).

19. There is no such thing as the experience of the *future*—all that we know is the *present*, so if there is to be time (other than now) there must be some evidence of the *past.*

20. See 'The Representative Theory Of Memory', in Paul Edwards, ed., 1972, *The Encyclopedia Of Philosophy*, vol. 5, p. 266.

   Also: "These attempts to conceive retrospection in general, and of memory in particular, in a non-dualistic way are, then, unsuccessful; when not equivalent to denying that we ever remember at all, they are equivalent to admitting that memory is a mode of representative knowledge ... And all that we call empirical knowledge consists primarily of memories, taken as true portrayals of past events as they were when they happened." (Arthur O. Lovejoy, 1955, *The Revolt Against Dualism*, pp. 383-4).

21. On a closely related subject a theoretical physicist has proposed the 'Past Hypothesis' is necessary in order to save the Big Bang Theory.

   "The truth is, we don't have any more direct empirical access to the past than we do to the future, unless we allow ourselves to assume a Past Hypothesis." (Sean Carroll, 2010, *From Eternity to Here*, p. 183).

   "Within our observable universe, the consistent increase of entropy and the corresponding arrow of time cannot be derived from the underlying reversible laws of physics alone. They require a boundary condition at the beginning of time. To understand why the Second Law works in our real world, it is not sufficient to simply apply statistical reasoning to the underlying laws of physics; we must also assume that the observable universe began in a state of very low entropy. David Albert has helpfully given this assumption a simple name: the *Past Hypothesis.*

   "The Past Hypothesis is the one profound exception to the Principle of Indifference that we alluded to above. The Principle of Indifference would have us imagine that, once we know a system is in some certain macrostate, we should consider every possible microstate within that macrostate to have an equal probability. This assumption turns out to do a great job of predicting the *Future* on the basis of statistical mechanics. But it would do a terrible job of reconstructing the *past*, if we really took it seriously." (Sean Carroll, 2010, *From Eternity to Here*, p. 176).

22. Many believe this is an area of research where future Nobel Prizes will be won.

23. It is arguably the case that digital computers are a great analog for studying the human brain. To be useful they must be designed to do what humans can do, cheaper or easier or better. Thus computers are designed by the human brain to replicate what the human brain does but cheaper, faster, better.

24. It is also generally agreed that focusing on one task at a time is more efficient than attempting to multi-task.

25. The discrepancy between what rises to consciousness and what is in the subconscious processing is five orders of magnitude in difference.
"... the human sensory system sends the brain about eleven million bits of information each second ... The actual amount of information we [consciously] can handle has been estimated to be somewhere between sixteen and fifty bits per second." (Leonard Mlodinow, 2012, *Subliminal: How Your Unconscious Mind Rules Your Behavior*, p. 33).

26. "General purpose computers all share the same architecture: the 'von Neumann Architecture', designed by John von Neumann in the 1950s." (Tadeusz Zawidzki, 2007, *Dennett*, p. 75).

27. Daniel Dennett likens this internal mental babble to a Joycean machine—the term Joycean comes from the 'stream of consciousness' style of narrative that James Joyce employed in some of his novels. The term 'stream of consciousness' originally comes from William James.

"Dennett understands the Joycean machine in terms of a metaphor drawn from computer science: it is a virtual machine running on the hardware of the brain. A virtual machine is a machine that is *simulated* on an actual computer, rather like virtual reality. Any general purpose computer, for example, the standard desktop personal computer, can implement numerous virtual machines." (Tadeusz Zawidzki, 2007, *Dennett*, p. 75).

28. "The illusion that there is a conscious self in the brain, first attending to one bit of information, then to another, is caused by this Joycean machine." (Tadeusz Zawidzki, 2007, *Dennett*, p. 75).

29. Chapter 8 will demonstrate that the *sixth-sense* as used in Eastern philosophy means *memory*.

30. "Instead of a fixed weekly cycle, they [the Umeda of the Borneo rain forest] make use of a set of seven words articulated to 'today', i.e. the day before the day before yesterday / the day before yesterday / yesterday / today / tomorrow / the day after tomorrow / the day after the day after tomorrow. "In the Umeda 'week', today, so to speak is always Wednesday." (Alfred Gell, 1992, *The Anthropology of Time*, p. 88).

31. Mihaly Csikszentmihalyi, 1990, writes about unmediated reality in *Flow, The Psychology of Optimal Experiences.*

32. Buddhism refers to 'greed, anger and ignorance' as the three evils and all are a consequence of the temporal realm ... Ignorance is thinking the self-ego is real, and thus becoming involved in the illusion of existing in time and space. Greed and anger are a consequence of thinking in this way. Greed is due to our striving for the (nonexistent) future, and anger results when our expectations are not met. There are other consequences, such as remorse for the past and fear of the future. Thinking temporally incites regret and fuels ambition and worry.

PART II: ANAMNESIS

Chapter Seven: Archaeology of the Self: Language, Duality, Belief

1. "In the Platonic theory of knowledge, the opposition between the plain of *Aletheia* and the plain of *Lethe* mythically represents the opposition between the act of anamnesis—an escape from time, a revelation of immutable and eternal being—and the error of *Lethe*—human ignorance and forgetfulness of the eternal truths." (Marcel Detienne, 1999, *Masters of Truth*, pp. 121-2).
2. The following chapter (8) will detail why the mysterious sixth sense from eastern philosophy is best defined as *memory*.
3. "The Greek word anamnesis means remembering or recollection and is the basis of Plato's theory of knowledge and wisdom." (Richard Geldard, 2000, *Remembering Heraclitus*, p. ix).

    "A philosopher, it appeared, had to engage in an anamnetic exploration of his own consciousness in order to discover its constitution by his own experience of reality, if he wanted to be critically aware of what he was doing ... It had to go as far back as his remembrance of things past would allow in order to reach the strata of reality-consciousness that were the least overlaid by later accretions. The *anamnesis* had to recapture the childhood experiences that let themselves be recaptured because they were living forces in the present constitution of his consciousness." (Eric Voegelin, 1990, *Anamnesis*, p. 12-13).

4. Gregory Nixon voices his suspicion that time-language-self are closely intertwined: "Language, subjective consciousness, and time may well be different aspects of the same underlying reality." (Gregory Nixon, in Francisco Varela and Jonathan Shear, eds., 1999, *The View From Within: First Person Approaches to the Study of Consciousness*, p. 262).

5. Hugh J. Silverman, in Richard Kearney, ed., 1997, *Routledge History of Philosophy*, vol. 8, p. 391.

6. "It belongs to the nature of language that it relates to nothing external." (Gunter Figal, in Robert J. Dostal, ed., 2002, *The Cambridge Companion to Gadamer*, p. 115).

7. "As Poincaré says, 'If all the assertions which mathematics puts forward can be derived from one another by formal logic, mathematics cannot amount to anything more than an immense tautology. Logical inference can teach us nothing essentially new, and if everything is to proceed from the principle of identity, everything must be reducible to it. But can we really allow that these theorems that fill so many books serve no other purpose than to say in a round-about fashion 'A=A'?'" (A. J. Ayer, 1952, *Language, Truth and Logic*, p. 85).

8. "One is to make the reasonable but solipsistic assumption that, because language cannot point outside itself, we must remain forever inscribed in its sign-circulation." (David Loy, 1988, *Nonduality*, p. 259).

    The condition of the hermeneutic circle is also thought to apply to history and past: "... all we have in history is a series of constructed texts commenting on constructed texts commenting on constructed texts, in a seemingly endless circle of constructed meanings which cannot be directly assessed against an unmediated 'real' past." (Mary Fulbrook, 2002, *Historical Theory*, p. 19).

9. "Does this mean that skepticism has won the day? Is the complex relativity of all our knowledge claims, which hermeneutical philosophers insist upon,

just the most recent version of skepticism? Some philosophers within the hermeneutical tradition (and some outside it, as well) think that this is indeed the case. In their eyes hermeneutics is just skepticism dressed up in French and German clothes." (Brice Wachterhauser, in Robert J. Dostal, ed., 2002, *The Cambridge Companion to Gadamer*, p. 69).

10. Richard A. Watson, in Robert Audi, ed., 1999, *The Cambridge Dictionary of Philosophy*, p. 244.

11. "In *The Quest for Certainty*, Dewey finds all the problems of modern philosophy derive from dualistic oppositions, particularly between spirit and nature." (Richard A. Watson, in Robert Audi, ed., 1999, *The Cambridge Dictionary of Philosophy*, p. 245).

12. Richard A. Watson, in Robert Audi, ed., 1999, *The Cambridge Dictionary of Philosophy*, p. 245.

13. Daniel Dennett, 2005, *Sweet Dreams: Philosophical Obstacles to a Science of Consciousness*, p. 3.

14. Richard A. Watson, in Robert Audi, ed., 1999, *The Cambridge Dictionary of Philosophy*, p. 245.

15. "In moral terms there is no good without evil, and in logical terms there is no truth without error. These, according to Hegel, are central truths of dialectics." (H. B. Acton, in Paul Edwards, ed., 1972, *The Encyclopedia Of Philosophy*, vol. 3, p. 245).

16. Richard P. Feynman, 1998, *The Meaning of It All: Thoughts of a Citizen Scientist*, p. 33.

17. Eric Voegelin, 1990, *Anamnesis*, pp. 12-13.

18. "In fact, belief and unbelief are strictly issues for the ego; you can't be an unbeliever unless there are some believers against whom you are an unbeliever. All such oppositions are creations of the ego." (James P. Carse, 1994, *Breakfast at the Victory: The Mysticism of Ordinary Experience*, p. 11).

19. "The difficulty is to realize the groundlessness of our believing." (Ludwig Wittgenstein, 1969, *On Certainty*, p. 24e).

20. Karl Popper, 1966, *The Open Society and Its Enemies*, vol. 2, p. 221.

21. Michel de Montaigne, 1987, *Apology for Raymond Sebond*, p. 53.

22. David Loy, 2009, *Awareness Bound and Unbound: Buddhist Essays*, p. 71.

23. Richard Brodie, 1996, *Virus of the Mind: The New Science of the Meme*, p. 32.

24. Will Durant, 1926, *The Story of Philosophy*, p. 102.

25. René Descartes, quoted in David J. Darling, 2004, *The Universal Book of Mathematics: From Abracadabra to Zeno's Paradoxes*, p. 90.

26. Dogen Zenji, 1975, *Shobogenzo*, vol. 1, p. 16.

Chapter Eight: Sixth Sense: Memory and Mind

1. "Mind is memory, at whatever level, by whatever name you call it; mind is the product of the past, it is founded on the past, which is memory, a conditioned state." (J. Krishnamurti, 1975, *The First And Last Freedom*, p. 209).

2. Hannah Arendt, in a footnote stated that the first use of the "sixth sense" when referring to the common sense was by Gottsched in 1730 (Hannah Arendt, 1997, *The Life of the Mind*, vol. 1, p. 221).
    At this same time in the East this term (sixth sense) had already been in use for at least two thousand years.

3. "What since Thomas Aquinas we call common sense, the *sensus communis*, is a kind of sixth sense needed to keep my five senses together ..." (Hannah Arendt, 1997, *The Life of the Mind*, vol. 1, p. 50).

4. David Loy, 'The Mahayana Deconstruction of Time', in *Phil East and West*, Jan. 1986, p. 13-23. This was reprinted in David Loy, 1988, *Nonduality*, p. 219.

"In the experience of enlightenment, the citta [mind] is said to be 'freed' from the 'point of view' that is the self."(Princeton Dictionary of Buddhism, p. 194).

5. What has been termed the Santiago School is an expressed acknowledgment that all we can know are our own personal sensations, and all else is fabricated from those sensations. "Maturana and Varela do not maintain that there is a void out there, out of which we create matter. There is a material world, but it does not have any predestined features. The authors of the Santiago theory do not assert that 'nothing exists'; they assert that 'no things exist' independent of the process of cognition. There are no objectively existing structures; there is no pregiven territory of which we can make a map—the map itself brings forth the territory." (Fritjof Capra, 1997, *Web of Life: A New Scientific Understanding of Living Systems*, p. 271).

6. Anthony R. McIntosh, in Henry L. Roediger, et al., eds., 1975, *Science of Memory: Concepts*, p. 59.

7. Dalai Lama XIV, (Tenzin Gyatso), 2000, *Dzogchen, The Heart Essence of Great Perfection,* p. 107.

8. "Why the Buddhists with their otherwise thorough analysis of the mental factors should have paid so scant attention to the phenomenon of memory of the past as such has remained a riddle that needs to be examined." (Padmanabh S. Jaini, in Janet Gyatso, ed., 1992, *In The Mirror Of Memory: Reflections on Mindfulness and Remembrance in Indian and Tibetan Buddhism*, p. 48).

9. Edward S. Casey, 1987, *Remembering : A Phenomenological Study*, p. 290.

10. Modern India and China are nearly as time-oriented as is Western culture. If there is an unbridgeable cultural gap, many modern Indian and Chinese are caught on the Western side of that gap.

11. "It is generally believed that one reason why East and West will never meet is that the Indians had no history until Greek historians taught them how to

mark off historical periods by dates and how to trace consequences to causes and so transform poetical and mythical accounts of the Indian past into histories, and that the Chinese who, although they had histories that recorded the past and clocks which measured the lapse of time, had no knowledge of the nature of time and developed no science of mechanics." (Richard McKeon, in Charles M. Sherover, ed., 1975, *The Human Experience of Time: The Development of its Philosophic Meaning*, p. 573).

12. Howard Trivers, 1985, *The Rhythm of Being: A Study of Temporality*, p. 225.

13. Hajime Nakamura, 'Time in Indian and Japanese Thought' in J. T. Fraser, ed., 1966, *The Voices of Time,* p. 81.

14. "Awareness is seeing the discovery of mindfulness ...The Sanskrit word for awareness is *smriti* which means 'recognition,' 'recollection.' Recollection not in the sense of remembering the past but in the sense of recognizing the product of mindfulness." (Chogyam Trungpa, 1988, *The Myth of Freedom: and the Way of Meditation*, p. 49-50).

15. D. T. Suzuki, 2000, *Outlines of Mahayana Buddhism*, p. 118.

Chapter Nine: Mind-Body, Self-Other, and the Explanatory Gap

1. Richard Noll and Carol Turkington, 1994, *Encyclopedia of Memory and Memory Disorders*, p. vii.

2. Bertrand Russell, 1959, *My Philosophical Development*, p. 26.
   "Whereas the metaphysical brand of solipsism has had no outright advocates, the epistemological brand has had many. In some degree or style it has been espoused by almost every major philosopher since Descartes. After all, it is easy to believe that knowledge about what exists is empirical and that all empirical knowledge originates in inner cognitive states of persons." (Paul Edwards, 1972, *The Encyclopedia Of Philosophy*, vol. 7, p. 490).

"I do not think this theory (solipsism) can be refuted, but I also do not think that anybody can sincerely believe it." (Bertrand Russell, 1959, *My Philosophical Development*, p. 104).

"As regards the world in general, both physical and mental, everything that we know of its intrinsic character is derived from the mental side, and almost everything that we know of its causal laws is derived from the physical side. But from the standpoint of philosophy the distinction between physical and mental is superficial and unreal." (Bertrand Russell, 1954, *The Analysis of Matter*, p. 402).

3. William James said of his concept of *Radical Empiricism*: "To be radical, an empiricism must not admit into its constructions any element that is not directly experienced, nor exclude from them any element that is directly experienced." (William James, 1971, *Essays in Radical Empiricism*, p. 25). He meant this to be inclusive of psychic experience.

    "Radical empiricism consists first of a postulate ... The postulate is that the only thing that shall be debatable among philosophers shall be things definable in terms drawn from experience." (C. W. Huntington with Geshe Namgyal Wangchen, quoting James, 1989, *Emptiness of Emptiness*, p. 44). The mind-body problem could never be understood empirically unless both mind and body are included in an empiricism that studies both the psychic and the physical. William James called this *radical empiricism*.

4. Kant felt there were two forms of "pure intuition", namely space and time, one for the outer sense and one for the inner. (Bertrand Russell, 1945, *A History of Western Philosophy*, pp. 708, 713).

5. Edmund Husserl, 1991, *On the Phenomenology of the Consciousness of Internal Time (1893–1917)*.

6. Robert Sokowolski, describing Edmund Husserl's philosophy, in Robert Audi, 1999, *The Cambridge Dictionary of Philosophy*, p. 406.

7. 'The explanatory gap' is attributed to Joseph Levine (1983), see John Horgan, 2000, *The Undiscovered Mind: How the Human Mind Defies Replication, Medication, and Explanation,* p. 16.

8. Donald M. Borchert, ed., 1996, *The Encyclopedia of Philosophy,* Supplement , p. 97, refers to the explanatory gap as the "heart of the Mind-Body problem".

9. "... the [hard problem/explanatory gap] problem of consciousness [is] one of the most exciting intellectual challenges of our time. Because consciousness is both so fundamental and so ill understood, a solution to the problem may profoundly affect our conception of the universe and of ourselves." (David J. Chalmers, 1996, *The Conscious Mind: In Search of a Fundamental Theory*, p. xii).

10. "The question that philosopher David Chalmers has seductively and successfully nicknamed 'The Hard Problem'... seems just as far from having an answer today (or, for that matter, at any time in the future) as it was many centuries ago." (Douglas Hofstadter, 2007, *I am a Strange Loop*, p. 361). "The Hard Problem (Chalmers 1995) is an extension of the Explanatory Gap." (Michael Tye, *2009, Consciousness Reconsidered: Materialism Without Phenomenal Concepts*, p. 144).

11. Ned Block, in Donald M. Borchert, ed., 1996, *The Encyclopedia Of Philosophy*, Supplement, p. 98.

12. "It is just that with the concepts we have and the concepts we are capable of forming, we are cognitively closed to a full, bridging explanation by the very structure of our minds. There is such an explanation, but it is necessarily beyond our cognitive grasp." (Michael Tye, *Mind* 108, pp. 705-725).

13. "In other words, I think the explanatory gap is constructed by the ways we think about these matters linguistically rather than by the underlying primary-process brain matters themselves." (Jaak Panksepp, in Shaun Gallager and Jonathan Shear, eds., 1999, *Models of the Self*, p.124).

14. "If all that matters is the computation, we can ignore the brain's wiring diagram, and its chemistry, and just worry about the 'software' that runs on it." (Daniel Dennett, 2005, *Sweet dreams: Philosophical Obstacles to a Science of Consciousness*, p. 18).

    Dennett points out that Julian Jaynes first advanced this sort of thinking: "We can only know in the nervous system what we have known in behavior first. Even if we had a complete wiring diagram of the nervous system, we still would not be able to answer the basic question. Though we knew the connections of every tickling thread of every single axon and dendrite in every species that ever existed, together with all its neurotransmitters and how they varied in its billions of synapses of every brain that ever existed, we could still never—*not ever*—from a knowledge of the brain alone know if that brain contained a consciousness like our own. We first have to start at the top, from some conception of what consciousness is, from what our own introspection is." (Julian Jaynes, 1975, *The Origin of Consciousness in the Breakdown of the Bicameral Mind*, p. 18).

15. Ned Block, speaking about the explanatory gap: "This is the heart of the Mind-Body Problem." (Ned Block, in Donald M. Borchert, ed., 1996, *The Encyclopedia Of Philosophy*, Supplement, p. 97).

16. Julian Jaynes, 1975, *The Origin of Consciousness in the Breakdown of the Bicameral Mind*.

17. Daniel Dennett, 1998, *Brainchildren: Essays on Designing Minds*, p. 129.

18. Ibid., p. 130.

    Not only does Dennett say that language was necessarily present, Jaynes stipulates that *time* was necessarily present: "My essential point here, however, is that history is impossible without the spatialization of time that is characteristic of consciousness." (Julian Jaynes, 1975, *The Origin of Consciousness in the Breakdown of the Bicameral Mind*, p. 251).

"In oral cultures, tradition is not known as such, [it doesn't seem like the past] even though these cultures are the most traditional of all. To understand tradition, as distinct from other modes of organizing action and experience, demands cutting into time-space in ways which are only possible with the invention of writing. Writing expands the level of time-space distanciation and creates a perspective of past."(Anthony Giddens, *The Consequences of Modernity*, p. 37).

19. Greek culture left the best historical record of the change in the interpretation of the sense of memory from meaning Truth (i.e. the present experience of memory) to representing the ego's past.

20. Karl Popper called the period that included the Buddha, Lao Tse, Socrates, and Confucius the *Axial Age*. Recently, Karen Armstrong called this period *The Great Transformation* in a book by that title. Armstrong and others have expanded the period to be the 1st millennium BC, but centered on the influence of these same people. This was a time when a greater sense of self-awareness came on the scene. Also see discussion on Axial Age in Chapter 2.

21. Primitive cultures have a primitive sense of time; as a consequence, there is less of an ego that is 'out-there'. Modern Western culture has a sophisticated time concept that can project a big ego. Time is first formally spatialized in the Axial Age and consciousness becomes more self-conscious. This is often mistaken for 'raising consciousness'.

Chapter Ten: Reason, Sense, and Modernity

1. Edward S. Casey, 1987, *Remembering: A Phenomenological Study*, p. 4.

2. Edith Hamilton and Huntington Cairns, eds., 1961, *The Collected Dialogues of Plato,* 'The Apology', 38a.

3. Brooks Haxton, trans., 2001, *The Collected Wisdom of Heraclitus*, p. 81.

4. Ludwig Wittgenstein, 1969, *On Certainty*, p. 23e.

5. René Descartes (1596–1650). His most important works on method: *Discourse on Method* (1637); *Meditations on First Philosophy* (1641); *Principles of Philosophy* (1644).

6. René Descartes. John Cottingham, ed. and trans., 1986, *Meditations on First Philosophy*, p. 12.

7. Karl Popper, 1985, *Unended Quest*, p. 87.

8. J. Krishnamurti, 1969, *Freedom From The Known*, p. 16.

9. René Descartes. John Cottingham, ed. and trans., 1986, *Meditations on First Philosophy*.

10. Russell pointed out that this didn't go far enough. Descartes should have said "there are thoughts", thus omitting the presumption of an I. (Bertrand Russell, 1945, *A History of Western Philosophy*, p. 567.)

11. God and Ego are dualistic in the sense that God depends on Ego's belief for existence, and Ego depends on God for giving meaning to existence. Ego without a god is trapped in the temporal world grasping for meaning through wealth, fame and power, all ending at death.

12. " ... horses would draw the forms of the gods like horses, and cattle like cattle, and they would make their bodies such as they had themselves." (Xenophanes (fifth century B.C.), in C. C. W. Taylor, ed., 1997, *Routledge History of Philosophy*, vol. 1, p. 71).

PART III: FINDING TIME

Chapter Eleven: Constructing Time—Time Reversal and Time Duality

1. Ludwig Boltzmann (1844–1906), quoted by Sklar, 1985, *Philosophy and Spacetime Physics*, p. 310.

2. Time reversal enters into physics because time is represented as algebraic quantities that can be represented either positively or negatively. Robert G. Sachs elaborates with the following technical explanation of time reversal: "How then does time reversal enter into physics? The answer is that it enters naturally as a result of the implicit assumption that the variable t, introduced to quantify the measure of time, is an algebraic variable; time intervals are additive. Time intervals are physically measurable quantities, and they can be assigned algebraic signs because we can add them to make longer intervals and subtract one from another to construct shorter intervals. Once the choice of an origin $t = 0$, for the time variable has been made, the assignment of a sign, let us say, positive for t later than $t = 0$ and negative for t earlier than $t = 0$ appears to be merely a matter of convention. The 'time reversed' variable $t' = -t$ appears to have equal standing because time intervals $\Delta t = t_2 - t_1$ can just as well be expressed in terms of t', $\Delta t' = t'_2 - t'_1$ without altering their essential algebraic properties." (Robert G. Sachs, 1987, *The Physics of Time Reversal*, p. 4).

3. " ... the laws of physics that have been articulated from Newton through Maxwell and Einstein, and up until today, show a complete symmetry between past and future." (Brian Greene, 2004, *The Fabric of the Cosmos: Space, Time, and the Texture of Reality*, pp. 144-5).

4. Paul Davies, 1995, *About Time*, p. 223.

    Huw Price also cites an interesting statement by Davies regarding time reversal when speaking of the present world in time reversed language: "Its occurrence is no more remarkable than what we at present experience—indeed it is what we actually experience—the difference in description being purely semantic and not physical." (Huw Price, 1996, *Time's Arrow and Archimedes' Point: New Directions for the Physics of Time*, p. 100.)

Richard Morris concedes the direction of time is not discernible. He cites time reversal due to a collapsing universe. "But if intelligent beings still existed at this point in the evolution of the universe, they would notice nothing unusual about any of this. Since the processes taking place in their brains would also be reversed, they would 'think backwards' and view phenomena exactly as we do." (Richard Morris, 1986, *Time's Arrow: Scientific Attitudes Toward Time*, p. 211).

Time Sutra asserts that the arrow of time is directly attributable to assuming whether *memory is past,* or conversely, *memory is future.* To assume the latter, is absurd, but no more logically absurd that assuming the former.

5. Roger Penrose is quoted saying that a reversal in entropy would cause a reversal in the perception of time (Huw Price, 1996, *Time's Arrow and Archimedes' Point: New Directions for the Physics of Time*, p. 102). Price also notes that Stephen Hawking said a reversal of entropy would cause a psychological time reversal (Ibid., p. 103).

6. If the particles are in time, then they are not random, they have a history. Entropy is founded on the assumption that since it is impossible to measure every particle's position and momentum, then randomness of position and momentum is assumed, but without any logical justification. When the history is lost, prediction becomes impossible.

7. "On reflection, Melhberg considered his scientific obsession to be the problem of time-reversal ... " Henry Mehlberg, 1980, *Time, Causality, and the Quantum Theory*, vol. 1, p. xvii.

8. Anisotropy (of time) refers to a time in which past and future can be distinguished from each other. Whereas, isotropic time (isotropy) means time symmetry—there is no discernable arrow of time.

9. Henry Mehlberg, 1980, *Time, Causality, and the Quantum Theory*, vol. 2, pp. 200-1.

Mehlberg concludes that time is symmetrical (isotropic) and insists that we shouldn't let the presumptions contained in language get in the way.

"To cut a long story short: these vital temporal words, like 'future', 'past', 'before', 'after', etc., are 'egocentric particulars' exactly like the spatial adverbs 'here', 'there', 'underneath', etc. The possibility and convenience of using egocentric particulars while referring to space never prevented anybody from considering space isotropic. There is neither more nor less to the allegedly disastrous isotropy of time." (Henry Mehlberg, 1980, *Time, Causality, and the Quantum Theory*, vol. 2, p. 202).

10. Mehlberg, 1980, *Time, Causality, and the Quantum Theory*, vol. 2, p. 164.

11. Huw Price, 1996, *Time's Arrow and Archimedes' Point: New Directions for the Physics of Time*, p. 158.

12. Victor J. Stenger, 2000, *Timeless Reality: Symmetry, Simplicity, and Multiple Universes*, p. 12.

13. J. T. Fraser, 1988, *Time: the Familiar Stranger*, p. 289.

    Fraser errs by omitting *philosophy* from his list of the places to look for the arrow of time.

14. J. B. Priestly, 1968, quoting M. F. Cleugh in *Man and Time*, p. 49. Cleugh credits McTaggart with recognizing that the appearance of time must still be explained (see M. F. Cleugh, 1937, *Time: and Its Importance in Modern Thought*, Methuen & Co. LTD London).

15. "Instead of a fixed weekly cycle, they (the Umeda) make use of a set of seven words articulated to 'today', i.e. the day before the day before yesterday / the day before yesterday / yesterday / today / tomorrow / the day after tomorrow / the day after the day after tomorrow. In the Umeda 'week', today, so to speak, is always Wednesday." (Alfred Gell, 1992, *The Anthropology of Time: Cultural Constructions and Temporal Maps and Images*, p. 88).

    The Umeda, in 1976 at the time of the publication, were a small hunter-gatherer society in the rainforest of Borneo.

16. Mircea Eliade has made this point throughout his writings: "In studying these traditional societies, one characteristic has especially struck us: it is their revolt against concrete, historical time, their nostalgia for periodical return to the mythical time of the beginning of things, to the 'Great Time'. (Mircea Eliade, 1954, *The Myth of the Eternal Return*, p. ix).

   "All sacrifices are performed at the same mythical instant of the beginning; through the paradox of rite, profane time and duration are suspended." (Mircea Eliade, 1954, *The Myth of the Eternal Return*, p. 35).

   "In other words, he (the shaman) succeeds in abolishing history (all the time that has elapsed since the 'fall' and the severance of direct communication between heaven and earth); he 'returns to the past', he re-enters the primordial paradisiac condition." (Mircea Eliade, 1952, *Images and Symbols*, p. 167).

   "We must never forget that one of the essential functions of the myth is its provision of an opening into the Great Time, a periodic re-entry into Time primordial." (Mircea Eliade, 1960, *Myth, Dreams, and Mysteries*, p. 34).

   "Every religious festival, any liturgical time, represents the reactualization of a sacred event that took place in a mythical past, 'in the beginning'." (Mircea Eliade, 1957, *The Sacred and the Profane*, p. 69).

17. Mihaly Csikszentmihalyi, 1990, *Flow: The Psychology of Optimal Experiences*, p. 121.

18. Alfred Gell, 1992, *The Anthropology of Time: Cultural Constructions of Temporal Maps and Images*, p. 22.

19. Jay Griffiths, 2004, *A Sideways Look At Time*, p. 65.

20. Mircea Eliade, 1963, *Myth and Reality*, p. 140. Although Eliade here speaks of archaic man, he said the same of all primitive cultures.

21. Samsara, Skt. (see glossary) a term used in the East to mean the illusory world in which humans are involved, and the *triple time* (Skt. *Trikala*) simply means the mundane 'world of the three times' of the past, present and future.

22. The reverse arrow of time can never have the psychological attraction that the promise of future has in a world with a forward arrow of time. The promises of future-time, for the naive believer, appears to be limited only by insufficient ambition and imagination.

23. Ludwig Wittgenstein, 1921, *Tractatus Logico-Philosophicus*, 6.4311.

Chapter Twelve: Linear Time, Cyclic Time, and Rhythm

1. Lewis Mumford, in David S. Landes, 1983, *Revolution in Time: Clocks and the Making of the Modern World,* p. 187.

2. The American social philosopher Lewis Mumford has said of the clock that it "disassociated time from human events and helped create the belief in an independent world of mathematically measured sequences: the special world of science." (Anthony F. Aveni, 1995, *Empires of Time*, p. 158).

3. Albert Einstein and Leopold Enfield, 1938, *The Evolution of Physics: From Early Concepts to Relativity and Quanta*, p. 295.

4. It has also been effectively argued that the second was derived from dividing the hour into 60 minutes and then dividing the minute into 60 seconds. This was tied to the Egyptian 12 hours of night and 12 hours of light (also included in Babylonian and Sumerian thought). Additionally, the influence of the zodiac and astrology were tied in with the 360 degrees of a circle which might have come from the approximately 360 days in the year. Taken together, these would give a geometric mathematical grounding that is loosely tied to the phenomena of the zodiac and seasons, but the close rhythmic correspondence between the second and the heartbeat would not have gone unnoticed.

5. The oscillation between day and night was problematic for early time keepers because the periods of light and dark varied by latitude and season. Whereas, the cesium atomic clock is so accurate it measures the decay in the earth's

spin due to the tidal drag of the moon, thus we have 'leap seconds' to correct for the increasing length of planetary rotation.

6. "I have said earlier that physics doesn't contain the idea of a present instant or of a time that passes. I must explain this statement, since if it is true, then physics will provide a way of describing our sensations, and to some extent even our consciousness, without the use of these ideas, and such a description may enable us to understand the myth. There is, in fact, no quantitative measure of the passage of time that can be used scientifically, and I cannot find any statement one can make about it that is capable of scientific proof. How quickly does time pass? At a rate of one second per second? That will get us nowhere." (David Park, in J. T. Fraser et al., eds., 1972, *The Study of Time*, p. 112).

7. 'Primitive man' is not a pejorative, it is intended to indicate 'living closer to nature'. This does not mean that primitive man is enlightened, but he recognizes his entrapment and designs a thousands cures through a thousand different ceremonies, taking place over a thousand different cultures. And in every instance, the intent of the ceremony is to remove time and return to the eternal present. Linear time of modern humans is conceived by ordering the recognized cycles along a continuum. This successive arrangement of cycles (*i.e. calendar*) results in a linear temporal perspective that is quite different from the primitive perspective.

8. John Hassard, in Samuel L. Macey, ed., 1994, *Encyclopedia of Time*, p. 170.

9. Alfred Gell, 1992, *The Anthropology of Time: Cultural Constructions of Temporal Maps and Images,* p. 4.

10. John Hassard, in Samuel L. Macey, ed., 1994, *Encyclopedia of Time,* p. 170.

11. E. R. Leach, 1961, *Rethinking Anthropology*, p. 126.

12. Alfred Gell, 1992, *The Anthropology of Time: Cultural Constructions of Temporal Maps and Images,* p. 31.

13. E. R. Leach, 1961, *Rethinking Anthropology*, p. 134.

14. Ibid., p. 135.

15. Rhythm has many potential synonyms but they all seem to implicate time: periodic, recurring, reverberation, oscillation, cyclical, pulsation, repetition, etc.

16. The quote continues: "Recent psychological tests seem to confirm a rhythmical organization in time perception. An article by Ernst Poeppel entitled 'Oscillations as Possible Basis for Time Perception' sets forth the evidence. Poeppel stated his conclusions as follows: '... the temporal continuum is subjectively quantitized into discrete units which successively follow each other'." (Howard Trivers, 1985, *The Rhythm of Being: A Study of Temporality*, pp. 102-3.)

17. Albert Einstein, 1982, *Ideas and Opinions*, p. 363.

18. Ibid.

19. Ibid..

20. Howard Trivers, 1985, *The Rhythm of Being: A Study of Temporality*, p. 23.

21. Society is most cohesive when we imitate each other, we adopt memes, and many, perhaps most, of the memes are socially beneficial. But there is one meme that is so deeply buried in our psyche that it lies undetected. This would not necessarily be a problem, except it happens to be the single most influential meme that humans adopt, and at a very early age. The result is that the adult human is time-bound, constrained to act in time, not understanding timeless nature.

22. There is a liver fluke (*Dicroelium dendriticum*) that in one stage of its life cycle infects and then takes control of the ant's behavior so that it will likely be eaten by cattle and complete its life cycle in the liver of one of the cattle. The ant is manipulated to leave the ant nest at dusk to climb to the tip of a stem of grass and clamp on with its mandibles until dawn, then, if not eaten, climb down and return to the nest till the next evening. Daniel Dennett has used a similar analogy comparing the infected ant with the human's tendency

251

to fall under the control of religious ideology, with the subsequent loss of awareness, and sometimes with disastrous consequences.

23. Eckhart Tolle, 1999, *The Power of Now: A Guide to Spiritual Enlightenment*, p. 50. The Buddhists have pared this list down to the *three evils*: Greed, Anger, and Ignorance.

Chapter 13: Time-Space Duality and the Block Universe

1. Dzogchen Ponlop, 2006, *Best Buddhist Writing 2006*, p. 235.

2. Humans have an innate proclivity to believe. We come pre-adapted to believe. Any animal that learns survival traits from its parent must come equipped to 'believe' what they are 'told'. There is an inherent interest in what the parent is doing, and those who pay attention increase the probability of surviving their youth. For the human animal, the parent invests years in teaching the child to talk and think, and those children who do not learn to talk and think never become fully functioning members of society. It is a matter of survival that we come preprogrammed to believe. This is the attitude needed to promulgate a culture of survival. But what to do when the cultural beliefs are threatening the culture's own survival? When culture enters a death spiral brought on by unquestioning belief, then only *doubt* can change thinking.

3. Koan: In Zen, a paradox to be meditated upon, and when resolved will lead to insight.

4. Jean Piaget, 1967, *The Child's Conception of Space,* p. 51.

5. Bertrand Russell, 1921, *The Analysis of Mind*, p. 47.

6. William James appears to be the first to use 'block universe' (Bertrand Russell, 1945, *A History of Western Philosophy,* p. 801).

7. Howard Trivers, 1985, *The Rhythm of Being: A Study of Temporality,* pp. 26-7.

8. J. R. Lucas, 'A Century of Time', in Jeremy Butterfield, ed., 1999, *The Arguments of Time,* p. 6.

9. Palle Yourgrau, 2005, *A World Without Time: The Forgotten Legacy of Gödel and Einstein,* p. 115.

10. Karl Popper, 1985, *Unended Quest,* p. 129.

11. Dan Falk, 2008, *In Search of Time: The Science of a Curious Dimension,* p. 170.

12. Maurice Nicoll, 1953, *Living Time,* p. 69.

PART IV: THINKING ABOUT TIME: THREE PHILOSOPHERS

Chapter 14: Derrida and Deconstruction

1. St. Augustine, *Confessions*, p. 223. Augustine goes on to say that past and future only "... exist in the mind, and I find them nowhere else ... "(ibid.)

2. One of the earliest attempts at reconciliation was by Immanuel Kant. "In his *Critique of Pure Reason* (1781), Kant sought a grand reconciliation, aiming to preserve key lessons of both rationalism and empiricism." (Robert Audi, ed., 1999, *The Cambridge Dictionary of Philosophy*, p. 273).
William James tried to integrate all the senses (body and mind) through his radical empiricism that understood if the mental is not included in any explanation of the world then it is an inadequate explanation of reality.

3. The dominant movement in rationalism in the latter half of the 20th century was the 'linguistic turn', superceding the existential movement, which faded

253

after mid-century. Jacques Derrida is representative of late 20th century development of rationalist philosophy—his early influence came from the existentialist writers and later, as a mature philosopher, he employed the arguments of linguistic theory.

4. Stephen Hawking, 1988, *A Brief History of Time: From the Big Bang to Black Holes*, pp. 174-5.

    Hawking is right to criticize the state of philosophical rationalism, but he seems unaware that all his mathematical constructions are also caught in the hermeneutic circle—the same self-referential trap in which language is caught.

5. Stefan Klein is referring to our unhealthy hectic lives in the following quote, "Perhaps you can judge the inner health of a land by the capacity of its people to do nothing." (Stefan Klein, quoting de Grazia, 2006, *The Secret Pulse of Time: Making Sense of Life's Scarcest Commodity*, p. 271).

6. Prometheus is sentenced to eternal torture for befriending humankind by giving humans fire. For many, fire is metaphor for the Greek *techne,* meaning the craft of technology.

7. "Existentialism, a philosophical and literary movement that came to prominence in Europe, particularly in France after World War II ..." (Robert Audi, ed., 1999, *The Cambridge Dictionary of Philosophy*, p. 296). This is the existential situation in which humans find themselves, thrown into an absurd world without meaning, nevertheless having to make a life out of it. Existentialism was dominantly a continental philosophy—the catastrophe of WWII, just one generation after the catastrophe of WWI, injected a sense of heightened urgency to find authentic existence in a meaningless world.

8. Existentialism's concern for nature was primarily directed at the nature of man and the nature of being. Concern for a naturalist-environmental awareness did not appear among the existentialist's more pressing concerns;

that was left mostly to the empiricist, whereas most existentialist thought was derived from rationalist thinking.

9. Some of the existentialist philosophers did seem to offer an answer along the lines of 'just do it' and 'be authentic'; Commit to action; Live your life; Throw yourself into *being*. But if your *being* is driven by bad software, advocating action seems like a bad idea.

   Susan Sontag describes Sartre's answer: "According to Sartre's phenomenology of action, to act is to change the world. Man, haunted by the world, acts. He acts in order to modify the world in view of an end, an ideal. An act is therefore intentional, not accidental, and an accident is not to be counted as an act." (Susan Sontag, 1961, *Against Interpretation and Other Essays*, p. 95).

10. Taken from the Jacques Derrida entry in *Wikipedia*.

11. Sartre said in *Nausea*, "The true nature of the present revealed itself: it was what exists, and all that was not present did not exist. The past did not exist." Cited in Ralph B. Winn, 1960, *Dictionary of Existentialism*, p. 75.

    "The insatiable for-itself, running after itself in an eternal and useless pursuit, is the source of time." (David Loy, paraphrasing Sartre, 1996, *Lack and Transcendence: The Problem of Death and Life in Psychotherapy, Existentialism, and Buddhism*, p. 36).

12. The lineage of influence leading up to Jacques Derrida (1930–2004) was Franz Brentano (1838–1917), Edmund Husserl (1859–1938), Martin Heidegger (1889–1976), and Jean-Paul Sartre (1905–1980). Derrida (1930–2004) is at the end of an existential lineage.

    This was a lineage concerned with time—all in this lineage doubted time, but all, in the end, reified time, including Derrida.

13. E. O. Wilson, 1998, *Consilience: The Unity of Knowledge*, p. 41. Of course, E. O. Wilson is very much an empiricist and he probably also had Derrida in mind when he said, "Postmodernism is the ultimate polar antithesis of the

255

Enlightenment. The difference between the two extremes can be expressed roughly as follows: Enlightenment believers think that we can know everything, and radical postmodernists believe we can know nothing." (E. O. Wilson, 1998, *Consilience: The Unity of Knowledge*, p. 40).

14. The *linguistic* turn was a major movement in 20th century Western philosophy that is characterized by philosophy turning its focus on language, based on the idea that the most fundamental and intractable problems of philosophy are related to the way we use language. Michael Dummett attributes the *linguistic turn* to German philosopher and mathematician Gottlob Frege (1848–1925), " ... due to Gottlob Frege, epistemology was supplanted by the philosophy of language as the fundamental field of philosophical investigation. Frege's reorientation of philosophy ... finally directed philosophers' attention at the proper focus: the relation of language to reality. Dummett is thus a leading advocate of the 'linguistic turn'." (Donald M. Borchert, ed., 1996, *The Encyclopedia Of Philosophy,* Supplement, p. 132).

Charles Pierce (1839–1914) is also cited as a founder "of philosophy's so-called linguistic turn". (Donald M. Borchert, ed., 1996, *The Encyclopedia Of Philosophy,* Supplement, p. 393).

15. " ... facts enabled Saussure to claim that language is to be largely understood as a closed formal system of differences" (Robert Audi, ed., 1999, *The Cambridge Dictionary of Philosophy*, p. 816).

16. From the Jacques Derrida entry in Wikipedia.

17. When the word on one side of a binary opposition (duality) is expressed, it naturally refers to the opposite side, from which it derives its meaning.

18. Eva Ruhnau, 'The Deconstruction of Time and the Emergence of Temporality', in Harald Atmanspacher, et al., eds., 1997, *Time, Temporality, Now: Experiencing Time and Concepts of Time in an Interdisciplinary Perspective*, p. 54.

19. David Wood, 1989, *The Deconstruction of Time,* p. 2, quote taken from Jacques Derrida's, 1972, *Ouisia et Gramme.*

20 Derrida's 'deconstruction' was taken from Heidegger's '*destruktion*'. (D. M. Borchert, ed., 1996, *The Encyclopedia Of Philosophy,* Supplement, p. 123). *Destruktion* plays a prominent role in Heidegger's *Being and Time.*

21. "The explanation for Derrida, lies in the fact that the idea of unmediated or perfect presence is a pervasive and hidden prejudice carried forward from ancient times." (Richard Kearney, ed., 1997, *Routledge History of Philosophy*, vol.8, p. 460).

"Since the present is now seen as a production out of past and future, Derrida draws the conclusion that the present is made up out of what is different from it. Since by the same token the present never really arrives, but is always delayed due to the role future plays in its production, Derrida names the interweaving that produces time 'differance,' which captures both the differing and deferring aspects. Since 'space' is what is different from time, Derrida paradoxically defines 'differance' as the 'becoming-time of space' and the 'becoming-space of time'." (Samual L. Macey, ed., 1994, *The Encyclopedia of Time*, p. 161).

Chapter 15: Nietzsche and Eternal Recurrence

1. Sean Carroll, quoting Nietzsche, 2010, *From Eternity to Here: The Quest for the Ultimate Theory of Time*, p. 366.

2. Karl Jaspers repeatedly points out the inconsistencies in Nietzsche: "One finds it insufferable that Nietzsche said first this, then that, then something entirely different." (Karl Jaspers, 1997, *Nietzsche: An Introduction to an Understanding of His Philosophical Activity*, p. xi).

"The original plan called for a chapter made up of quotations from Nietzsche's naturalistic and extremist pronouncements, collected as evidence of his aberrations. The result was devastating, and I omitted it out of respect for Nietzsche." (Karl Jaspers, 1997, *Nietzsche: An Introduction to an Understanding of His Philosophical Activity*, p. xiii).

"All statements seem to be annulled by other statements. *Self-contradiction* is the fundamental ingredient in Nietzsche's thought. For nearly every single one of Nietzsche's judgements, one can also find an opposite. He gives the impression of having two opinions about everything. Consequently it is possible to quote Nietzsche at will in support of anything that one has in mind. Most parties have been able, at some opportune moment, to invoke Nietzsche: conservatives and revolutionaries, socialists and individualists (as well as those who are indifferent to politics), methodical scientists and idealistic dreamers, atheists and believers, freethinkers and fanatics. Consequently many have concluded that Nietzsche is full of confusion, is never in earnest, abandons himself to his own whims, and that it does not pay to take his inconsequential chatter seriously." (Karl Jaspers, 1997, *Nietzsche: An Introduction to an Understanding of His Philosophical Activity*, p. 10). He darts from one side of the duality to the other: "He takes up his position with the directness of a vital force and in a tone of absolute assertion. It is as though he has just arrived at the sole authentic truth. Then the process of doubting and the transition to the opposite pole is carried out with the same vigor." (Karl Jaspers, 1997, *Nietzsche: An Introduction to an Understanding of His Philosophical Activity*, p. 390).

"To proceed logically is to show that Nietzsche's pronouncements are *self-contradictory.*" (Karl Jaspers, 1997, *Nietzsche: An Introduction to an Understanding of His Philosophical Activity*, p. 418).

"He never realizes that he is wanting in the sort of philosophical training that comes through a painstaking study of the great thinkers. It is understandable that such things remain inessential to him since he himself stands with ever increasing decisiveness at the sources of philosophizing. His own originality allows him to take his professional inadequacies lightly by ignoring them. But his lack of restraint with regard to contradictoriness, accompanied by a tendency to allow his understanding (Verstand) to indulge in undialectical forms of thought—forms which he in turn scorns—constitutes a serious formal flaw in his entire work and an obstacle to an understanding of him." (Karl Jaspers, 1997, *Nietzsche: An Introduction to an Understanding of His Philosophical Activity*, pp. 418-19).

Harold Bloom complicates any interpretation even further, " ... Nietzsche ... has to be read antithetically, as meaning the reverse of what it appears to say ... " (Harold Bloom, 2004, *Where Shall Wisdom Be Found?*, p. 228).

3.   The following are some statements about salvation from time:

"Deliverance from this world and attainment of salvation are tantamount to a deliverance from cosmic Time." (Mircea Eliade, in H. S. Prasad, 1992, *Time in Indian Philosophy: A Collection of Essays*, p. 110).

"Our salvation consists in seeing the world *sub specie aeternitatus*, as God sees it, and in gaining thereby freedom from the bondage of time." (Scruton,1999, *Spinoza*, p. 30).

" ... that double headed monster of damnation and salvation—Time." (Samuel Beckett, 1999, *Proust and Three Dialogues with Georges Duthuit*, p. 11).

"The place and function of sacrifice is precisely here. The essence of sacrifice is not the lamb, the flower, the bread or a particular rubric, but the ritual through which Man finds salvation, specifically, overcomes the dominion of time and is rescued from its slavery." (Raimundo Panikkar, in J. T. Fraser, et al., eds., 1978, *The Study of Time III*, p. 709).

"Memory's powers are twofold. As a religious power, it is the water of Life, which marks the end of the cycle of 'metensomatoses'; as an intellectual faculty, it constitutes the discipline of salvation that results in victory over time and death and makes it possible to acquire the most complete kind of knowledge.

"According to the dichotomatic view of the philosophicoreligious sects, earthly life was corrupted by time, which was synonymous with death and oblivion. Man was cast into the world of *Lethe*, where he wandered in the meadow of *Ate*." (Marcel Detienne, 1999, *Masters of Truth in Archaic Greece*, pp. 122-3).

"Supernatural theology, by falsifying man's nature and linking his salvation to the illusory notions of God and immortality, has entirely subverted ethical truth." (Paul-Henri Holbach's ethics in Paul Edwards, ed., 1972, *The Encyclopedia of Philosophy*, vol. 4, p. 50).

"In this sense Samkhya-Yoga is a soteriology of the present. Salvation is attained in the present ... " (Braj M. Sinha, 1983, *Time and Temporality in Samkhya-Yoga and Abhidharma Buddhism*, p 149).

"At the moment when individual salvation becomes a preoccupation, man seeks to achieve it by integrating himself with the whole. What he asks of memory is not knowledge of his own past but the means to escape time and be reunited with the divine." (Jean-Pierre Vernant, *2006, Myth and Thought Among the Greeks*, p. 136).

4. The *Birth of Tragedy* (1872) was published when Nietzsche was 28 years old, and already he was concerned with the past. Nietzsche saw memory as "it was", meaning 'the past'. Stambaugh explaining Nietzsche: "Man lost the natural creature's relationship to its past when he learned to understand the 'it was'." (Joan Stambaugh, 1987, *The Problem of Time in Nietzsche,* p. 46).

"Nietzsche is able to say that existence is an uninterrupted having been." (Joan Stambaugh, 1987, *The Problem of Time in Nietzsche*, p. 46).

Stambaugh goes on: "In each moment the animal is completely what it is. Man is never entirely what he is, but encounters only what he was. But the strange thing in respect to his past is that it is not annihilated by the moment that follows. It is not annihilated at all, but somehow acquires its own independent power. Man finds himself threatened by this power of the past and asks which attitude toward the past is appropriate *vis-a-vis* life. Man lost the natural creature's relationship to its past when he learned to understand the 'it was'." (Joan Stambaugh, 1987, *The Problem of Time in Nietzsche*, p. 46).

5. Will Durant, 1926, *The Story of Philosophy*, p. 314.

6. Nietzsche borrowed the "eternal return" from the Greek stoics. (Frederick Copleston, 1993, *A History of Philosophy*, vol. 1, p. 389).
Nietzsche most likely took his endlessly cyclic world from the stoics. (J. O. Urmson and Jonathan Ree, eds., 2000, *The Concise Encyclopedia of Western Philosophy and Philosophers*, p. 248).

7. "The conception of the will to power is central in Nietzsche's philosophy." (Paul Edwards, ed., 1972, *The Encyclopedia of Phil.*; vol. 5, p. 510).

8. Joan Stambaugh, 1987, *The Problem of Time in Nietzsche*, p. 46.
This is Stambaugh's doctoral dissertation, which offers a comprehensive overview of how Nietzsche used time throughout his various writings. She points out some of the self-contradictions involved.

9. "And if I do not succeed to such an extent that people render their highest vows in my name for thousands of years, then, in my estimation, I shall have accomplished nothing." (Karl Jaspers, quoting Nietzsche, 1997, in *Nietzsche: An Introduction to an Understanding of His Philosophical Activity*, p. 413).
"Nietzsche's passionate extension of his egoism into the realm of metaphysics leads to more confusion than even his rhetorical gifts were able to hide. Moreover a philosopher who said, 'there are no truths, only interpretations', risks the retort: 'Is that true, or only an interpretation?'" (Scruton, on Nietzsche, 1995, *A Short History of Modern Philosophy: From Descartes to Wittgenstein*, p. 198).

" ... to strive after a little immortality—I have never been modest enough to demand less of myself. The aphorism, the apophthegm, in which I am the first master among Germans are the forms of 'eternity'; my ambition is to say in ten sentences what everyone else said in a book—what everyone else *does not say* in a book ... I have given mankind the profoundest book it possesses, my *Zarathustra*..." (Nietzsche, R. J. Hollingdale, trans., 1990, *Twilight of the Idols*, p. 115).

"It is true that expressions such as the following appear unambiguously to support the unlimited absoluteness of the individual: 'Selflessness is worthless.' 'One must stand firmly upon his own feet.' 'Sanctify the ego.' The 'creating, willing, evaluating ego' is 'the measure and value of things.' 'Will to be a self!' But these are to be set over against the following: His 'ponderous, serious, granite ego' says to itself: 'What do I matter!' And with respect to man, we read that, 'we are buds on a single tree ... the individual per se is a delusion ...You must put an end to feeling yourself as such a phantastic ego!'" (Karl Jaspers, 1997, *Nietzsche: An Introduction to an Understanding of His Philosophical Activity*, p. 423).

10. "In 1867 Nietzsche, now a student at Leipzig, was treated by two Leipzig doctors for a syphilitic infection; but there existed no cure for syphilis and the disease took its course." (Bernd Magnus and Kathleen M. Higgins, eds., 1996, *Cambridge Companion to Nietzsche,* p. 80).

11. "Unamuno dismissed Nietzsche's eternal recurrence as 'a sorry counterfeit of immortality'." (David Loy, 1996, *Lack and Transcendence: The Problem of Death and Life in Psychotherapy, Existentialism, and Buddhism*, p. 120). Will Durant said further: "The idea of eternal recurrence, thought common to the 'Apollonian' Spencer as well as to the 'Dionysian' Nietzsche, strikes one as unhealthy fancy, a weird last minute effort to recover the belief in immortality." (Will Durant, 1926, *The Story of Philosophy*, p. 331).

12. "In July and August [1881] came the time which he was to remember until the end as having given birth to one of his ideas that he considered most profound [eternal recurrence]" (Karl Jaspers, 1997, *Nietzsche: An Introduction to an Understanding of His Philosophical Activity*, p. 48).

Never mind that he had obviously come across this idea in his Greek studies.

13. "Mircea Eliade rightly observed that this doctrine is 'a manifestation of ontology uncontaminated by time and becoming.' Cyclical time, in which the future has already been past and the past will be future an infinite number of times, is no time at all." (Milič Čapek, in David Ray Griffin, ed., 1986, *Physics and the Ultimate Significance of Time*, p. 300).

"If the recurrence [eternal return] happens without my being conscious of it as repetition, it can have no significance to me. If, on the other hand, I am aware of it as a repetition, what 'recurs', is by virtue of that consciousness, not exactly what originally occurred." (Genevieve Lloyd, 1993, *Being in Time: Selves and Narrative in Philosophy and Literature*, p. 110).

14. The physical sciences present a very logical argument for complete determinism—free will is an assumption that runs counter to empirical observation. Recent observations in the cognitive sciences suggest that the sense of free will is not real, but only seems that way. Neurological experiments, first performed by Benjamin Libet, have demonstrated that what appear to be the intentional actions of the self only rise to consciousness after the action has been unconsciously initiated (see Benjamin Libet, 2004, *Mind Time: The Temporal Factor in Consciousness*).

"He [Libet] is famous in part for discovering that we unconsciously decide to act well before we think we've made the decision to act. This finding has major implications for one of the deepest problems in philosophy and psychology, namely the problem of 'free will.'" (Benjamin Libet, from the foreword by S. M. Kosslyn, 2004, *Mind Time: The Temporal Factor in Consciousness*, p. x).

Recent research involving fMRI imaging in real-time has added to the evidence that action is determined prior to consciousness of the decision to act.

15. Nietzsche, *The Birth of Tragedy*, cited in Joan Stambaugh, 1987, *The Trouble With Time in Nietzsche,* p. 43.

Nietzsche is wrong. The animal doesn't forget, it just doesn't get its memories entangled in the time concept.

16. Joan Stambaugh, 1987, *The Problem of Time in Nietzsche*, p. 45.

17. Joan Stambaugh, quoting Nietzsche, 1987, *The Problem of Time in Nietzsche*, p. 44.

Nietzsche continues: "And yet the child's play must be disturbed: only too soon will it be called out of its forgetfulness. Then it comes to understand the phrase 'it was', that password with which struggle, suffering and boredom approach man to remind him what his existence basically is—a never to be completed imperfectum (Imperfektum)." (Joan Stambaugh, 1987, *The Problem of Time in Nietzsche*, p. 44).

18. Ibid., p. 57.

19. Friedrich Nietzsche, 1997, *Untimely Meditations,* p. 66.

20. Joan Stambaugh, 1987, *The Problem of Time in Nietzsche*, p.59

Another translation of the same passage by R. J. Hollingdale: "One is reminded almost palpably of the very relative nature of all concepts of time: it almost seems as though many things belong together and time is only a cloud which makes it hard for our eyes to perceive the fact." (Friedrich Nietzsche, 1997, *Untimely Meditations*, p. 209).

21. Joan Stambaugh, quoting Nietzsche, 1987, *The Problem of Time in Nietzsche*, p. 59.

22. Friedrich Nietzsche, 1997, *Untimely Meditations,* p. 66.

23. Karl Jaspers, 1997, *Nietzsche: An Introduction to an Understanding of His Philosophical Activity*, p. 351.

24. Ibid., p. 350. *amor fati*: loving one's fate.

25. Stambaugh insists that 'will to power' cannot be separated from Nietzsche's concept of time. (Joan Stambaugh, 1987, *The Problem of Time In Nietzsche*, p. 92)

26. Nietzsche is considered one of the early thinkers who came to be known as an Existentialist. "Existentialism: a chiefly 20th century philosophical movement embracing diverse doctrines but centering on analysis of individual existence in an unfathomable universe and the plight of the individual who must assume ultimate responsibility for his acts of free will without any certain knowledge of what is right or wrong or good or bad." (*Merriam-Webster's Collegiate Dictionary*).

27. "It also exposes as legend the story of Nietzsche as the great mind 'driven insane' by solitude, lack of understanding, and the vulgarity of the world around him ... " (Bernd Magnus and Kathleen M. Higgins, eds., 1996, *Cambridge Companion to Nietzsche*, p. 80).

Chapter 16: Longchenpa

1. Juan Mascaro, trans., 1962, *Bhagavad Gita*, p. 53. This is said in the voice of Krishna, who also said: "I am all powerful Time which destroys all things ..." (ibid. p. 92).

2. Graham Coleman, ed., 1994, *Handbook of Tibetan Culture: A Guide to Tibetan Centres and Resources Throughout the World*, p. 335.

3. All the following Longchenpa quotes are taken from the various texts of the *Seven Treasures*.

4. "**Dzogchen**: Tib., lit., "great perfection", the primary teaching of the Nyingmapa school of Tibetan Buddhism. This teaching, also known as ati-yoga (extraordinary yoga), is considered by its adherents as the definitive and most secret teaching of Shakyamuni Buddha ... According to the experience of *dzogchen* practitioners, purity of mind is always present and needs only to be recognized." (Stephan Schumacher and Gert Woerner, eds., 1989, *Encyclopedia of Eastern Philosophy and Religion*, p. 97).

5. Chokyi Nyima Rinpoche, Erik Hein Schmidt, trans., 1989, *The Union of Mahamudra and Dzogchen*, p. 248. Other variations: "Cutting the stream of the thoughts of the three times." (Tsele Natsok Rangdrol, Erik Hein Kunsang, trans., *1989, Lamp of Mahamudra*, p. 80.) "Cutting through the stream of the delusion of time ... " (Padmasambhava, Erik Hein Kunsang, trans., 1990, *Dakini Teachings: Padmasambhava's Oral Instructions to Lady Tsogyal*, pp. 185-6.).

6. Longchenpa uses the term 'naturally occurring timeless awareness' throughout the text of the *Way of Abiding*. The term is used to indicate the natural enlightened experience of the world.

7. "But one thing is very clear that no Buddhist school holds time as a substantive reality ..." (H.S. Prasad, ed., 1991, *Essays on Time in Buddhism*, p. xi).

    "Ever since its origin, the Buddhist doctrine is concerned with the problem of time." (André Bareau, in H. S. Prasad, ed., 1991, *Essays on Time in Buddhism*, p. 1).

    " ... if the history of Indian Time philosophy is ever written, it will be in a large measure a history of Buddhist thought." (Stanislaw Schayer, in H. S. Prasad, ed., 1991, *Essays on Time in Buddhism*, p. 322).

8. Longchenpa, Keith Dowman trans., 2010, *Natural Perfection: Longchenpa's Radical Dzogchen*, p. 62. According to Barron's trans. this is a quote from the Tantra of the *Heaped Jewels*.

9. Longchenpa, Richard Barron trans., 2001, *A Treasure Trove of Scriptural Transmission: A Commentary on the Basic Space of Phenomena*. p. 222.

10. Ibid., p. 282.

11. Ibid., p. 147.

12. Ibid., p. 233.

13. Ibid., p. 362.

14. Ibid., p. 422.

15. Longchenpa, Richard Barron trans., 1998, *Way Of Abiding*, p. 127.

16. Ibid., p. 129.

17. Ibid., p. 17.

18. Longchenpa, Keith Dowman trans., 2010, *Natural Perfection: Longchenpa's Radical Dzogchen*, p. 5.

19. Naturally occurring timeless awareness is "a state of imperturbable rest not cultivated in meditation" (Longchenpa, Richard Barron trans., 1998, *Way Of Abiding*, p. 108).

    "timeless awareness (is) not cultivated in meditation" (Longchenpa, Richard Barron trans., 1998, *Way Of Abiding*, p. 181). "to abide in some vague (meditative) state is samsara, pure and simple." (Longchenpa, Richard Barron trans., 2001, *A Treasure Trove of Scriptural Transmission: A Commentary on the Basic Space of Phenomena*, p. 266).

20. Longchenpa, Kennard Lipman and Merrill Peterson trans., 2010, *You Are the Eyes of the World*, p. 37.

    In follow-up commentary on Longchenpa's text, Kennard Lipman said that it is incumbent upon us to learn our own nature: "It must be remembered that, as taught in Buddhism, compassion is not just love, kindness, and caring for the welfare of others, but it is also a knowledge of the fundamental causes of unhappiness and of the definitive means for uprooting them. Thus, without a knowledge of your natural state, compassion will always be contrived."

(Longchenpa, Kennard Lipman and Merrill Peterson trans., 2010, *You Are the Eyes of the World*, p.72).

21. Nagarjuna (2nd/3rd century CE), One of the most important philosophers in Buddhism.

22. Longchenpa, Richard Barron trans., 1998, *Way Of Abiding*, p. 11.

23. In Samuel L. Macey's *Time: A Bibliographic Guide* (1991) he estimated in 1989 that in the 20th century " ... about 95,000 books have been published on time-related subjects." (p. xviii).

Macey then notes that a full third of the books published relating to time, since the beginning of the 20th century, were published during the 7 years 1983–1989, just prior to the publication of his bibliographic guide. Up through the end of this last millennia the publishing frenzy pertaining to time continued, at the time of this writing (2013), publishers Blackwell and Oxford are coming out with large single volume references on *time* late this year.

PART V: THE PATH

Chapter 17: Abyss

1. Karl Jaspers, 1997, *Nietzsche: An Introduction to an Understanding of His Philosophical Activity*, p. 408.

2. We humans are not our self-egos. The self-ego is a virus that is written into memory by our culture, and designed to reinterpret memory as past.

3. Norbert Weiner, 1948, *Cybernetics*: the science of communication and control theory that is concerned especially with the comparative study of

automatic control systems (such as the nervous system and brain and mechanical-electrical communication systems).

4. Norbert Weiner, 1950, *The Human Use of Human Beings*, p. 138.

5. John Blofeld, trans., 1958, *The Zen Teachings of Huang Po*, p. 32.

6. Maya is a common term in Eastern philosophy and religion that refers to the illusion or deception that is inherent in our common view of the world.

7. William James, 'The Varieties of Religious Experience' in *William James: Writings 1902–10*, p. 347. James here is quoting J. A. Symonds who also experiences the "obliteration of space, time, sensation and the multitudinous factors of experience which seem to qualify what we are pleased to call our Self."

8. Thomas Cleary speaks of the abyss in terms of Great Death: "This last minute psychic flurry is a common theme in Buddhist iconography and mythology, because it actually does happen to almost everyone. Many people in deep meditation do become agitated or frightened at the eleventh hour, on the verge of the 'Great Death' of the false self, thus spoiling the result." (Thomas Cleary, 1993, *No Barrier: The Munmonkan*, pp. 205-6).

    "De Martino states that the Great Death entails at once Great Birth beyond birth and death and that the one to whom that happened in one sense ceases to 'be'." (Joan Stambaugh, 1999, *The Formless Self*, p. 64).

    And from the Advaita Vedanta tradition, Nisargadatta said, "Time will come to an end. This is called the Great Death (*mahamrityu*), the death of time." (Nisargadatta Maharaj, 1973, *I Am That*, p. 361).

9. *Postmodern* is a very theory-laden term, more specifically, a time-laden term. Modernity is dependent upon the theory of past. Post-modernity is dependent upon the idea that modernity is now passed and postmodernism is the new present. The term *postmodern* seems legitimate in architecture, but it seems absurd in philosophy. Philosophical postmodernism is a failure to understand modernity, which is a failure to understand time.

10. Alexandra David-Neel and Lama Yongden, 1967, *The Secret Oral Teachings In Tibetan Buddhist Sects*, p. 96.

11. David Loy, quoting Po-shan (Huang Po), 1988, *Nonduality*, p. 207.

Chapter 18: Modernity's End and The Path of Doubt

1. Nisargadatta Maharaj, 1973, *I Am That*, p. 495. Nisargadatta (1897–1981) was an influential Indian teacher and philosopher of Advaita Vedanta.

2. Thomas Kuhn, 1962, *The Structure of Scientific Revolutions*.

3. Morris Berman, 1990, *Coming To Our Senses: Body and Spirit in the Hidden History of the West*, p. 304.

4. "In the Dzogchen teachings there are practices ... to suit all kinds of birds and all kinds of cages. But one must know for oneself what kind of bird one is, and what kind of cage one is in. And then, one must really want to come out of all cages because it's no good just making one's cage a little bigger ... It's no good building a new crystal cage out of the Dzogchen teachings." (Namkhai Norbu, 2000, *The Crystal and the Way of Light: Sutra, Tantra, and Dzogchen*, p. 114).

5. Paul Davies, 1995, *About Time*, quoting David Bohm, p. 199. Davies goes on to say "History shows he is right. So often, major progress in science comes when the orthodox paradigm clashes with a new set of ideas or some new piece of experimental evidence that won't fit into the prevailing theories. Then somebody discards a cherished assumption, perhaps one that has almost been taken for granted and not explicitly stated, and suddenly all is transformed. A new more successful paradigm is born." The originator of quantum theory, Max Planck (1858–1947) famously said: "Science advances one funeral at a time."

6. Modernity, by its very name, means time. It invokes time, because it assumes the past. Modern means the present, and is defined by its contrast to

past—that which is not modern. We might imagine Derrida saying: "what is *not* being said when the word *modern* is invoked, is that which is *not* modern—the past." The modern is contingent on there being a past, without which modernity is impossible. Modernity is defined as 'not past'. Modernity/past is an obvious temporal duality. One side is dependent upon the other.

This creates an interesting linguistic argument that brings modernity to an end by removing its dualistic opposite, 'the past'. *Modern* means *present* with relation to the *past*. Without the past, *modernity* has no meaning. This is why I say that to bring modernity to an end, we must reason all the way to the end, but it can be as simple as recognizing the fact that, when we are no longer assuming past, we bring modernity to an end. Again, the key is discovering that we generally hold an unacknowledged theory about memory (the representative theory of memory); specifically this theory is founded on the premise *memory is past*. Remove this premise and the modern/past duality collapses. The idea of modernity can no longer be a part of our thoughts—remember, Ferdinand de Saussure said that words gain their meaning from their difference from other words. The most important word that is different from *modern*, is *past*. Without a past, the concept of modern loses all meaning.

7.   Quoting R. G. Collingwood, in Paul Edwards, ed., 1972, *The Encyclopedia of Philosophy*, vol. 2, p. 141.

8.   Francisco Sanches wrote his great treatise, *That Nothing Is Known*, in 1581: "Written at the same time that his cousin, Michel de Montaigne, wrote the *Apology For Raimond Sebond*, it devastatingly criticized the Aristotelian theory of knowledge." (*Cambridge Dictionary of Philosophy*, p. 810). The same article notes that Francisco Sanches was the first to use the term 'scientific method'.

9. "Some have claimed that Sanches was the first modern philosopher, preceding Descartes in the method of doubt and Bacon in the reliance on experience as the source of knowledge." (Paul Edwards, ed., 1972, *The Encyclopedia of Philosophy*, vol. 7, p. 279). Descartes must have learned of Sanches's thoughts and had probably read *That Nothing Is Known*.

10. Francisco Sanches, 1988, *That Nothing Is Known*, p. 173.

11. Ibid., p. 236.

12. Albert Camus, 1970, *Lyrical and Critical Essays*, p. 150.

13. This is the opening line of Bodhidharma's brilliant essay 'Outline of Practice' from Red Pine, trans., 1989, *The Zen Teachings of Bodhidharma*, p. 3.

14. David Ross Komito, trans., 1987, *Nagarjuna's Seventy Stanzas: A Buddhist Psychology of Emptiness*, p. 61.

15. If by *nihilism* we mean the annihilation of the ego, then indeed Buddhism could be considered nihilistic, but if the ego is unreal then this can't be taken as a criticism of Buddhism, rather it is a compliment. However, if nihilism means a plunge into *nothingness*, then this is not a description of Buddhism.

16. "And the joke is that it is the ego, the 'me', that wants enlightenment and enlightenment cannot come until the 'me' is demolished. That is the paradox." (Ramesh S. Balsekar, 1992, *Consciousness Speaks*, p. 342). Balsekar is speaking from the Vedanta tradition of India.

17. Zen probably became particularly susceptible to this extreme nihilism because to escape all the cultural encrustation of hundreds of years, Bodhidharma brought Indian Buddhism to China without a lot of the cultural baggage. Bodhidharma's Buddhism shed the cultural accretion from Indian sutras (Zen is sometime called the teaching outside the teachings). But this left *Ch'an* (Chinese Zen), without a voice, there was nothing to say—many have misunderstood this and plunged into nihilism. Those who take this fall into nihilism are sometimes called Stone Buddhas.

18. Gary Snyder, Carole Tonkinson, ed., 1995, *Big Sky Mind: Buddhism and the Beat Generation*, p. 179.

19. "The philosopher must remain solitary, because this is what he *is* according to his nature. His solitude is not to be *admired*. Isolation is nothing to be wished for as such." (Martin Heidegger, Ted Sadler trans., 2002, *The Essence of Truth*, p. 63).

    "Almost everyone I've heard of for whom this nameless thing appears to have been genuine seems to have gone into a long gestation period ... ten, twelve, twenty years before any 'coming out.'... Even Hui Neng, the Sixth Zen Patriarch went and hid in the mountains for fifteen years after it happened." (James P. Carse, 1986, *Perfect Brilliant Stillness: Beyond the Individual Self*, p. 100).

20. The plural of Buddha is often used in Buddhist literature. Gautama Buddha was the historical Buddha, but Buddhism teaches that there are an infinite number of buddhas going both backward and forward in time.

21. Satori: a state of intuitive illumination sought in Zen Buddhism (*Merriam-Webster*).

22. Muso Soseki, 1989, *Sun At Midnight*, p. 147.

23. Thomas Cleary trans., quoting Hui-neng, 1997, *Kensho: Several Works by Haukuin*, p. vii.

24. Joseph Campbell, 1968, *Creative Mythology: The Masks of God*, p. 624.

25. Keizan, Thomas Cleary trans., 1990, *Transmission of Light: Zen in the Art of Enlightenment*, p. 18.

26. Chogyam Trungpa, 2001, *Glimpses of the Abhidharma*, p. 23.

27. The term Perennial Philosophy was first made prominent by Leibniz (1646–1716). Aldous Huxley describes Perennial Philosophy in a book by the same title. He said, "Rudiments of the Perennial Philosophy may be found among the traditionary lore of primitive peoples in every region of the

world, and in its fully developed forms it has a place in every one of the higher religions." (Aldous Huxley, 1945, *The Perennial Philosophy*, p. vii). Huxley traces the Perennial Philosophy back to the Axial Age in both the East and the West.

28. The purpose of silencing the mind in sitting meditation is to quiet the mind so that it can look into the structure of thought. Humans think constantly—it would be good if they thought about the nature of thinking. This is the ground of the perennial philosophy. *Time Sutra* proposes a modern perennial philosophy—same as all other perennial philosophies (they all dispense with time and self)—but *Time Sutra* describes the time-self relation in the language of modernity.

Chapter 19: Meditation on Doubt

1. Dogen, Taigen Dan Leighton and Shohaku Okumura, trans., 2004, *Dogen's Extensive Record: A Translation of the Eihei Koroku*, p. 618.

2. Robert Audi, ed., 1999, *The Cambridge Dictionary of Philosophy*, p. 760. Panagiotis Thanassas remarks that in the 6th century BCE Xenophanes was already espousing skepticism: "Xenophanes preferred a skeptical stance that denied human beings any possibility of reliable knowledge." (Panagiotis Thanassas, 2007 *Parmenides, Cosmos, and Being: A Philosophical Interpretation*, p. 87).
    And Parmenides was clearly advocating a radical skepticism 150 years prior to Pyrrho.

3. "Based on the advice given by the Buddha to His disciples, the primary recommendation that the Masters give to neophytes is: 'Doubt!'" (Alexandra David-Neel and Lama Yongden, 1967, *The Secret Oral Teachings In Tibetan Buddhist Sects*, p. 15).

4. Karl Jaspers, on Lao Tzu, 1966, *The Great Philosophers: Anaxamander, Heraclitus, Parmenides, Plotinus, Lao-Tzu, Nagarjuna*, p. 97.

   The quote taken from Lao Tzu's *Tao Te Ching* continues: "To pretend to know when you do not know is a disease. Only when one recognizes this disease as a disease can one be free of the disease."

5. "[Gottlob Ernst] Schultze declares that the skeptic's sole authority is reason and that the highest excellence of man consists in the perfection of this faculty. Schultze is indeed at pains to stress that skepticism is not at odds with reason, but is reason's only consistent position." (Frederick C. Beiser, 1987, *The Fate of Reason: German Philosophy from Kant to Fichte*, p. 270). "According to [Solomon] Maimon, the skeptic is a philosopher whose primary interest is the truth rather than the formal virtues of a principle." (Frederick C. Beiser, 1987, *The Fate of Reason: German Philosophy from Kant to Fichte*, p. 288).

6. "From the mid-19th century 'natural philosopher' was gradually replaced by 'scientist'." (William F. Bynum, ed., 1989, *Dictionary of the History of Science*, p. 287).

   The earliest citation using the word scientist in the Oxford English Dictionary is 1840, but many of the great scientists of the late 19th century still thought of themselves as naturalists. The earliest use of scientist (1833) was intended as a satirical comment on the state of natural philosophy by William Whewell (1794–1866).

7. "Philosophers are not the only nor the chief advocates of the unreality of time. Some of them, indeed, have expressed the view more explicitly than have scientists, but for centuries it has been the physicists themselves who have, in one way or another, attempted to abolish time, or, at least, becoming ... They have persistently sought to formulate laws which would remain invariant despite assumed reversal of the time order, and so, in effect, eliminate time." (Errol E. Harris, 1988, *The Reality of Time*, p. 42).

8. "Time ... gives nothing to see. It is at the very least the element of invisibility itself. It withdraws whatever could be given to be seen. It itself withdraws from visibility. One can only be blind to time, to the essential disappearance of time even as, nevertheless, in a certain matter, nothing appears that does not require and take time." (Elizabeth Grosz, quoting Derrida, 1999, *Becomings: Explorations in Time, Memories, and Futures,* p. 1).

9. "Derrida's attack on 'the privilege granted to the present' should not distract us from realizing that his own conception of time constitutes another version of the everyday and 'commonsense' conception of time." (David Loy, 1988, *Nonduality*, p. 255).

10. In modern physics this is the boundary between Einsteinian physics and quantum mechanics. On one side of this boundary is precise measurement, and on the other, the world takes on a probabilistic appearance. This point at which measurement is no longer accurate, is the point of the Heisenberg Uncertainty Principle.

11. For a more detailed explanation of how Derrida reasons from linguistic analysis to a theory of time see chapter 14, *Derrida and Deconstruction* in Part IV, *Thinking About Time: Three Philosophers.*

12. "Skepticism is a practical philosophy concerned with leading us to a state of tranquility (ataraxia). The Skeptic's method is to oppose arguments to arguments, or arguments to appearances, so as to lead us to suspend judgement on the propositions involved ... Skeptics do not, however, doubt the appearances themselves, so long as they are acknowledged to be appearances. Consequently, they acknowledge time's appearances, but they wish us to suspend any positive judgements regarding the nature and existence of time." (Phillip Turetzky, *Time*, 1998, p. 34).

    "Suspending judgement led to a state of mind called 'ataraxia', quietude, peace of mind, or unperturbedness." (Robert Audi, 1999, *The Cambridge Dictionary of Philosophy*, p. 851).

13. "But Timon makes clear that the key to Pyrrho's Skepticism, and a major source of his impact, was the ethical goal he sought to achieve: by training himself to disregard all perception and values, he hoped to attain mental tranquility." (Robert Audi, 1999, *The Cambridge Dictionary of Philosophy*, p. 760).

14. "[Sextus Empiricus] tells us several times that it is the skeptic's *acceptance* of the conclusion that nothing is by nature good or bad that produces the desired state of tranquility." (Sextus Empiricus, Richard Bett, trans. and ed., 2005, *Against The Logicians*, p. xxii).

15. Joseph Goldstein, *Tricycle*, Winter 2001, p. 71.

PART VI: CONSEQUENCES OF TIME

Chapter 20: Our Inner Zombie

1. W. Somerset Maugham, 1944, *The Razor's Edge*, p. 269.

2. The argument is at least as old as Descartes, who assumed that the human body along with all other animals were simply mechanisms, except the human attained consciousness through the possession of a soul. This is the source of the Cartesian mind-body duality.

3. Incidentally, Daniel Dennett finds the argument for zombies unconvincing and refers to this debate as an "Embarrassment of zombies" and asks "Must we talk about zombies?" Daniel Dennett, 2005, *Sweet Dreams*, p. 13.

4. "Also you need to be careful. Watch for hungry ghosts who look like people, exactly same as people, but not real people." (George Crane, quoting Tsung Tsai, 2005, *Beyond the House of the False Lama*, p. 127).

Philosopher Jacob Needleman, speaking about an experience on a crowded downtown street in a major U.S. city, said, "Most of the people I was seeing, in the inner state they were in at that moment, were not really people at all. Most were what the Tibetans call 'hungry ghosts.'" (Jacob Needleman, 2003, *Time and the Soul*, p. 10).

5. Ibid., p. 10.

6. Titus Carus Lucretius, Rolfe Humphries, trans., 1968, *The Way Things Are*, p. 116.

    Heraclitus (c. 540BCE- c. 480BCE) said something similar more than 400 years before Lucretius (c. 99BCE- c. 55BCE): "But as to the rest, they fail to notice what they do after they wake up, just as they forget what they do when they sleep." (Richard Geldard, 2000, *Remembering Heraclitus*, p. 81).

7. Quoting Dennett: the "conscious self" [is] "the program that runs on your brain's computer." (in Zoltan Torey, 2009, *The Crucible of Consciousness: An Integrated Theory of the Mind and the Brain*, p. 156).

8. Tadeusz Zawidzki, 2007, *Dennett*, p. 83.

9. Dennett cites an example of how we are prevented from questioning: "If (some) religions are culturally evolved parasites, we can expect them to be insidiously well designed to conceal their true nature from their hosts, since this is an adaptation that would further their own spread." (Daniel Dennett, 2007, *Breaking the Spell: Religion as a Natural Phenomenon*, p. 85).

10. John Blofeld, 1958, *The Zen Teachings of Huang Po*, p. 121.

11. Daniel Dennett, 2007, *Breaking the Spell: Religion as a Natural Phenomenon*, p. 207.

12. The term *Buddha* is Sanskrit for *awakened one*.

Chapter 21: Strange Loops, Dualities, and Memes

1. Nagarjuna, K. Venkata Ramanan, trans., 1966, *Nagarjuna's Philosophy: As Presented in The Maha–Prajnaparamita–Sastra*; p. 199.

2. Roberto Mangbabiera Unger, 2007, *The Self Awakened: Pragmatism Unbound*, p. 103.

3. The extended quote is as follows: "Our mathematical and logical reasoning perpetually suggest to us the reality of a timeless world. We are tempted to mistake this embalmed world for the real thing. However, nothing is more real than time. In a sense it is the only real thing." (Roberto Mangbabiera Unger, 2007, *The Self Awakened: Pragmatism Unbound*, p. 103.).

4. Roberto Mangbabiera Unger, 2007, *The Self Awakened: Pragmatism Unbound*, p. 103.

5. Unger also said: "However, it is not the perennial philosophy alone that resists recognizing the reality of time. The logical or mathematical relations among propositions, even when they refer to events that seem to take place in time, seem themselves timeless. Thus, after ridding ourselves of the perennial philosophy, we may continue to find a conspiracy against recognition of the reality of time established in the inner citadel of our mental life." (Roberto Mangbabiera Unger, 2007, *The Self Awakened: Pragmatism Unbound*, p. 81).

6. Roberto Mangbabiera Unger, 2007, *The Self Awakened: Pragmatism Unbound*, p. 103.

7. Kant felt that the five senses are of the body and are senses of space. The sense of time, however, is of the mind. Kant believed that this is the natural way in which humans perceive reality. He thought that we really were missing the most fundamental reality, which he called the "thing in itself" (*ding an sich*).

Perhaps the following quotes will help elucidate Kant's concept of reality, the *ding an sich*: Jeffrey Gray asks, "What is the real external world or *ding an sich* ... that lies beyond the constructions of the conscious mind?" (Jeffrey Gray, 2004, *Consciousness: creeping up on the hard problem*, p. 62).

"Were it necessarily the case that the world's intelligibility is ontologically constituted by our linguistically constituted epistemic endeavors, then what we would know in principle would not be an 'independent' world, but something like an image of ourselves or a self-projection mirrored back to us by something like a Kantian *Ding-an-sich*." (Brice Wachterhauser, in Robert J. Dostal, ed., 2002, *The Cambridge Companion to Gadamer*, p. 73).

"Our whole experience of the world, he [Kant] declared, is subject to three laws and conditions, the inviolable forms in which all our knowledge is effectuated. These are time, space and causality. But they are not definitions of the world as it may be in and for itself, of *Das Ding an sich*, independently of our apperception of it; rather they belong only to its appearance, in that they are nothing but the forms of our knowledge. All variation, all becoming and passing away is only possible through these three." (Thomas Mann, 1939, *Schopenhauer*, pp. 5-6).

8. "The fact is, we all more or less take for granted this notion of "Cartesian Ego" in our daily lives; it is built into our common sense, into our languages, and into our cultural backgrounds as profoundly, as tacitly, as seamlessly, and as invisibly as is the notion that time passes or the notion that things that move preserve their identity." (Douglas Hofstadter, 2007, *I am a Strange Loop*, p. 306).

9. "Ceasing to believe altogether in the "I" is in fact impossible, because it is indispensable for survival. Like it or not, we humans are stuck for good with this myth."(Douglas Hofstadter, 2007, *I am a Strange Loop*, p. 294).

10. Douglas Hofstadter, 2007, *I Am A Strange Loop,* p. xiii., quoted from the 1999 twentieth anniversary edition of *Gödel, Escher, Bach: An Eternal Golden Braid.*

11. Hofstadter has also published his ideas in Uriah Kriegel and Kenneth Williford, eds., 2006, *Self-Representational Approaches to Consciousness*, pp. 465-515.

12. "The TV loop is not a *strange* loop—it is just a feedback loop." (Douglas Hofstadter, 2007, *I am a Strange Loop*, p. 187).

13. "In any strange loop that gives rise to human selfhood, by contrast, the level-shifting acts of perception, abstraction, and categorization are central, indispensable elements. It is the upward leap from *raw stimuli* to *symbols* that imbues the loop with "strangeness". (Douglas Hofstadter, 2007, *I am a Strange Loop*, p. 187).

14. Ibid., p. 276.

15. "Our 'I's are self-reinforcing illusions that are an inevitable by-product of strange loops, which are themselves an inevitable by-product of symbol-possessing brains that guide bodies through the dangerous straits and treacherous waters of life." (Douglas Hofstadter, 2007, *I am a Strange Loop*, p. 292).

16. Ibid., p. 360.

17. A *hermeneutic circle* is a closed system of self-referential parts (or symbols) that exhibit a circularity of interpretation such that every part gains its meaning from other parts within the system. This circle of interpretation cannot be escaped. The hermeneutic circle is defined and discussed in Chapter 7.

18. This was Derrida's insight—by careful examination of the text you can discover what is not being said, which he claimed might be as important as what was being stated.

19. "Twentieth-century hermeneutics advanced by Heidegger and Gadamer radicalize this notion of the hermeneutic circle, seeing it as a feature of all knowledge and activity. Hermeneutics is then no longer the method of human sciences but 'universal', and interpretation is part of the finite and situated character of all human knowing." (Robert Audi, ed., 1999, *The Cambridge Dictionary of Philosophy*; p. 378).

20. The extended quote reads: "Indeed the fact that true faith doesn't need evidence is held up as its greatest virtue; this was the point of my quoting the story of Doubting Thomas, the only really admirable member of the twelve apostles." (Richard Dawkins, 1976, *The Selfish Gene*, p. 330).

21. Ibid., p. 330.

22. The science of *memetics* is discussed in chapter 6. "Memetics provides a new way of looking at the self. The self is a vast memeplex—perhaps the most insidious and pervasive memeplex of all ... (it) permeates all our experience and all our thinking so that we are unable to see it clearly for what it is—a bunch of memes." (Susan Blackmore, 1999, *The Meme Machine*, p. 231).

23. "Given the importance of cultural learning in human evolution, our minds, unlike those of other animals, are largely products of culture. This fact leads Dennett and others to defend a specific model of cultural evolution: the memetic model. The idea first proposed by Dawkins (1976), is that ideas passed down through culture, called 'memes', behave much like genes passed down through biological reproduction." (Tadeusz Zawidzki, 2007, *Dennett*, p. 83).

24. "Memes are skills, habits, songs, stories, or any other kind of information that is copied from person to person." (Richard L. Gregory, ed., 2004, *Oxford Companion To The Mind*, p. 558).

    "Any skill or set of information that catches on with human beings counts as a meme." (Tadeusz Zawidzki, 2007, *Dennett*, p. 83).

25. "Dawkins (1993) coined the term 'viruses of the mind' to apply to such memeplexes as religions and cults—which spread themselves through vast populations of people by using all kinds of clever copying tricks, and can have disastrous consequences for those infected." (Susan Blackmore, 1999, *The Meme Machine*, p. 22).

    "Memes for blind faith have their own ruthless ways of propagating themselves. This is true of patriotic and political as well as religious blind faith." (Richard Dawkins, 1976, *The Selfish Gene*, p. 198).

26. "The computers in which memes live are human brains." (Richard Dawkins, 1976, *The Selfish Gene*, p. 197).

27. Richard L. Gregory ed., 2004, *Oxford Companion To The Mind*, p. 558.

PART VII: RESOLVING TIME

Chapter 22: The Western Buddha

1. Brooks Haxton, trans., 2001, *The Collected Wisdom of Heraclitus*, p. 51, Fr. 80.

2. James William Coleman, 2001, *The New Buddhism: The Western Transformation of an Ancient Tradition*, p. 54.

3. Mu Soeng, 2004, *Trust in Mind: The Rebellion of Chinese Zen*, pp.vii-viii.

4. Donald Lopez (*Tricycle: The Buddhist Review*, Fall 2002, p.47) makes a good case that there is already a Western Buddhism which he calls Modern Buddhism. It was "begun in part as a response to the *threat* of modernity" (p.114), and encroaching Christian dogma. Modern Buddhism has taken up numerous Western values, and in so doing has become acceptable to mainstream Western culture.

5. James William Coleman, 2001, *The New Buddhism: The Western Transformation of an Ancient Tradition*, p. 60.

6. The Beat Generation's influence on American Buddhism has become widely recognized. *Big Sky Mind: Buddhism and the Beat Generation* is an anthology of some of the Beat writers and poets of that time. *How the Swans Came to the Lake: a Narrative History of Buddhism in America* also documents the Beat influence on Buddhism.

7. James William Coleman, 2001, *The New Buddhism: The Western Transformation of an Ancient Tradition*, p. 62.

8. Melvin McLeod, ed., 2011, *Best Buddhist Writing 2011,* p. iv.

9. Stephen Batchelor, in Alan Hunt Badiner, ed., 2002, *Zig Zag Zen: Buddhism and Psychedelics*, p. 9.

10. Mu Soeng, 2005, in *Tricycle: The Buddhist Review*, Winter 2005, p. 49.

11. Sarvepalli Radhakrishnan and Charles A. Moore, eds., 1957, *A Sourcebook in Indian Philosophy*, p. 349.

12. Ryokan, Ryuichi Abe and Peter Haskell, trans., 1996, *Great Fool: Zen Master Ryokan—Poems, Letters, and Other Writing*, p. 45.

13. James William Coleman, 2001, *The New Buddhism: The Western Transformation of an Ancient Tradition*, p. 213.

14. David Loy, 1986, 'The Mahayana Deconstruction of Time', in *Phil East and West*, Jan. 1986, p. 17.

15. Thomas Kuhn's 1962 publication *The Structure of Scientific Revolutions,* considered one of the most influential texts of the 20th century, describes how Western science has evolved historically. Central to his history of science is the idea of *paradigms*—these are the philosophical and theoretical frameworks of a scientific discipline, which prescribe how *normal science* (Kuhn's term) is to be carried out within the frame of the paradigm. Change occurs when the paradigm is overturned and replaced by a new paradigm. How all this happens is the subject of his book. Ruling paradigm was not

Kuhn's term, but came from other authors remarking on his concepts of *paradigms* and the current *ruling paradigm theory of science*.

16. Present day Westerners who practice Buddhism often end up mistakenly trying to learn a deconstruction of a distant and ancient society, rather than deconstructing the present Western paradigm that dictates how we interpret the world.

17. The term *dharma* is used in a lot of ways, but here the intended meaning is "The teaching of the Buddha, who recognized and formulated this 'law'; thus the teaching that expresses the universal truth. The dharma in this sense existed already before the birth of the historical Buddha, who is no more than a manifestation of it." (Stephan Schumacher and Gert Woerner, Eds., 1989, *Encyclopedia of Eastern Philosophy and Religion*, p. 87).

18. "Buddhism is 2,500 years old and any thought system of that vintage has time to develop layers and layers of doctrine and ritual. Nevertheless, the fundamental attitude of Buddhism is intensely empirical and antiauthoritarian." (Henapola Gunaratana, 1992, *Mindfulness in Plain English*, p. 38).

19. These techniques appear in the literature, applied by some of the influential masters of the past—in specific, singular instances, these techniques worked well, but over time this has given rise to many 'masters' who seem unable to convey what they know into words. Even koan study (meditating on a paradoxical question proposed by the teacher) has become codified into a regimen of koans with a regimen of 'correct answers'.

20. Buddhism, by another name, is the Perennial Philosophy, but the greatest body of extant literature in the Perennial Philosophy comes from the many schools of Buddhism.

21. The mature Western Buddhism *describes the disease, explains how the contamination occurs, and prescribes a remedy*, which is identical to the

Buddha's recognition that there is suffering, there is a cause for suffering, and there is a path to end suffering.

Chapter 23: Allegories of the Labyrinth

1. Raimundo Panikkar, quoting Bhartrhari's *Vakyapadiya*, in H. S. Prasad, 1992, *Time in Indian Philosophy: A Collection of Essays*, p. 21.

2. Karl Jaspers comments on this, "*The turning around*: Human insight requires a turning around (*metastrophe*, *perigoge*) ... as in the cave the turning of the eyes involves the whole body, so knowledge, in turning from the realm of becoming to the realm of being, must take the whole mind with it." (Karl Jaspers, 1957, *Plato and Augustine*, p. 33).

   Charles Taylor explaining the "cave allegory": "Rather we see ourselves as having something like a capacity of vision which is forever unimpaired, and the move from illusion to wisdom is to be likened to our turning the soul's eye around to face in the right direction. Some people, the lovers of sights and sound and beautiful spectacles, are focused entirely on the bodily and the changing. Making these people wise is a matter of turning the soul's *gaze* from the darkness to the brightness of true being." (Charles Taylor, 1989, *Sources of the Self*, p. 123).

   The metaphor is also common in Eastern philosophy to "turn the gaze around" and look within—this is where enlightenment will be found. Padmasambhava, who conveyed Buddhism to Tibet from India in the 9th century said that if you want to know truth, "Seek, therefore, thine own Wisdom within thee." He also said, if you question the teachings, "To know whether this be so or not, look within thine own mind." (W. Y. Evans-Wentz, 2000, *The Tibetan Book of the Great Liberation*, pp. 238, 215).

3. Ironically, Socrates himself was eventually killed by these same prisoners (the Athenians) that he tried to free from their dark cavern of ignorance. Was

286

the Cave Allegory, a retelling by Plato of the circumstances of Socrates' death? Or was this a story that Socrates told, knowingly or unknowingly, that was a premonition of his own death?

4. Robert Audi, 1999, *The Cambridge Dictionary of Philosophy*, p. 710, summarizes Plato's opinion, "When we learn, we are really *recollecting* what we once knew and forgot."

5. Borges wrote *Library of Babel* in 1941, and published it in 1944 in *Ficciones,* a volume of short stories. He worked as an assistant librarian at this time in a small library in Buenos Aires (1937–1946), but was later the director of the Argentine National Library (1955–1973).

6. Actually, the library is self-referential, because it is contained within the bounds of language. The bounds of language are vast, but not infinite. The librarians in the Babel Library understand that the Library contains a finite number of books, and so the library itself is finite.

7. "When it was announced that the Library contained all books, the first reaction was unbounded joy." (Jorges Luis Borges, 2000, *The Library of Babel*, pp. 25-6).

   "At that same period there was also the hope that the fundamental mysteries of mankind—the origin of the Library and of time—might be revealed." (Jorges Luis Borges, 2000, *The Library of Babel*, p. 26).

8. Modern Information Theory would specify that "the ignorance, the trivial and the nonsensical" are only *noise*, and only the sought-after Truth would qualify as "information or data," which is referred to as the *signal*. In other words, the *signal to noise ratio* is so low that most information has been lost in the noise of the Internet.

9. We have a lifetime limit to our reading, which at best is a few thousand books. Read three books a week for 60 years and this is still less than 10,000 books—a small fraction of the new titles published just this year. There are millions of volumes on library shelves. Where do we begin?

287

10. "The beginning cannot be preserved as beginning; it can only be remembered or forgotten." (Richard I. Velkley, re. Heidegger's understanding of Aletheia, 2002, *Being After Rousseau*, p. 149.)

11. For a thorough treatment of the art of memory see Frances A. Yates, 1966, *The Art of Memory.*

12. Richard Noll and Carol Turkington, 1994, *Encyclopedia of Memory and Memory Disorders*, p. 150.

13. Jean Piaget, 1971, *A Child's Conception of Time*, p. 272.

14. Those who study the nature of memory notice two basic categories of long-term memory; memories that evoke a time/ and or place, which has come to be called the *episodic memory* (also called the autobiographical memory), and all the other memories that are not stamped with a time or place are referred to as semantic memory.

    "semantic memory: long-term memory for facts, other than autobiographical." (Richard L. Gregory, ed., 2004, *Oxford Companion to the Mind*, p. xviii).

15. "An animal or a newborn baby will experience a scene in reference to a self but will have no nameable self that is differentiated from within. Such a nameable self emerges in humans as higher-order consciousness develops during the elaboration of semantic and linguistic capabilities and social interactions." (Gerald Edelman, 2004, *Wider Than the Sky, The Phenomenal Gift of Consciousness*, p.73).

    "Tulving suggested that episodic memory, which requires conscious recollection of the time and place of some personal experience, is particularly characteristic of humans, whereas semantic memory, being the simple storage of a fact rather than a personal experience, is within the capacity of many animals." (Joseph LeDoux, 2002, *Synaptic Self: How Brains Become Who We Are*, p. 108).

16. M. G. F. Martin, in Hoerl and McCormack, eds., 2001, *Time and Memory: Issues in Philosophy and Psychology*, p. 260.

17. *Clew* is a term that can be traced back to the Sanskrit (*Merriam-Webster*) and it means a ball of yarn, and also means *clue*. And *clue* is defined as "something that guides through an intricate procedure or maze of difficulties." (*Merriam-Webster*).

18. The Perennial Philosophy is first recorded historically in the thinking of the Axial Age philosophers, but has recurred in many cultures throughout history. The term was first used by Agostino Steuco (1497–1548). Gottfried Wilhelm Leibnitz (1646–1716) also incorporated the term in his thinking.

19. Again, the apt metaphor taken from Wittgenstein; we are the fly, thinking our way out of the fly-bottle.

EPILOGUE

1. Michel de Montaigne, 1987, *An Apology for Raymond Sebond*, p. 170. Montaigne also wrote: "Knowledge begins with them (the senses) and can be reduced to them." (ibid., p.170).

2. "A Pali *sutta* that has become central in this regard is the *Kalama Sutta*, in which the Buddha exhorts a particular audience (notably, not his own disciples) not to believe any teaching because of tradition, scripture, or devotion to a teacher but to test its ideas for themselves ..." (David L. McMahan, *The Making of Buddhist Modernism*, p. 64).

   "The Buddha himself said to his monks: 'Just as the wise accept gold after testing it by heating, cutting, and rubbing it, so are my words to be accepted after examining them, but not out of respect for me.'" (Alan Wallace, *Dreaming Yourself Awake*, p. 69).

3. This is termed *lucid dreaming*, which is the awareness of the fact that one is dreaming, when one is still involved in the dream.

4. In human culture the ground is always defined as *memory is past*. I suspect that primitive cultures, if they possess an effective philosophico-religious system, will also have some strong attributes of the Perennial Philosophy.

5. What you want to hear can't be said, the only way that you can hear it, or in any way perceive it, is by looking for one's self. This goes beyond quieting the mind, you must awaken and look for yourself. At some point, and that point will always be 'here and now', realize how the temporal ego has taken complete control of the way we experience the world. *Time Sutra* is an argument that claims we can escape this way of thinking, and escaping from this thinking is the advent of enlightenment. It shows that there is a way, but the self must always make this discovery on its own.

6. Richard L. Velkley, 2002, *Being After Rousseau*, p. 149. I am repeating this quote because it is imperative that we remember.

7. *Time Sutra* offers a first principle in the form of an assumption that is universal for culture, which is, *memory is past*. I emphasize this as the key supposition that differentiates Samsara from Nirvana. On one side, the side of supposition, lies the Ego, whereas on the supposition-less side lies the Buddha. Crossing over is timeless, but while still caught in time, it seems endless, or worse, inevitably coming to an ending in time. The trajectory of Modernity, whose beginning is dated to the time of Bacon and Descartes, is completed in *Time Sutra*. Descartes's philosophical path of *doubt,* allied with the *empirical* bent of Bacon, created the intellectual atmosphere that launched Modernity. *Time Sutra* brings Modernity to its logical end by revealing the overarching conclusion of modernity that *reality is timeless.*

8. Meditation interrupts the ego-control, and takes the ego-program off-line for an hour. And those who meditate and knock the ego off-line for an hour, find meditation to be a success. And it is, because Being, even for a short time,

without ego, is nirvana. But when you stand up from the meditation cushion, there is a reversion to being in samsara. The hope of sitting meditation practice is that the ego-less thinking experienced on the cushion can be spread into the everyday world. After all, this is what Buddhism promises, the realization of our own Buddha Nature.

9. *Time Sutra* has the simplest mnemonic device: remember only that we have assumed *memory is past*. Observation confirms memory is present.

10. "What radicals need most right now isn't action but theory ... " Thomas de Zengotita, Common Ground: Finding Our Way Back to the Enlightenment, *Harper's*, Jan. 2003.

I think *Time Sutra* is an expression of the theory that Thomas de Zengotita was looking for.

Aletheia: *Aletheia* at the time of Parmenides' writing was a word that meant *Truth*, but was also the name for the Goddess of Memory, and translated literally the *aletheia* means *not-forgetting*. The literal meaning comes from the *'A'* of Aletheia meaning *not,* and *'lethe'* meaning *forgetting*, thus *Aletheia* literally means *not-forgetting*. In Parmenides' time, the term was used as a personification of the Goddess of Memory. The goddess who narrates Parmenides' poem is unnamed in the surviving text but it seems obvious that she is *Aletheia*, "Like *mnemosyne, Aletheia* is the gift of second sight: an omniscience, like memory, encompassing the past, present, and future." (Detienne, *Masters of Truth in Archaic Greece*, p. 65).

Anamnesis: Calling to mind knowledge that has been lost. "The Greek word anamnesis means remembering or recollection and is the basis of Plato's theory of knowledge and wisdom." (Geldard, *Remembering Heraclitus*, p. ix).

Anisotropy: Exhibiting properties with different values when measured in different directions. In reference to time anisotropy means that the past and future can be distinguished from each other. Whereas, isotropic time (isotropy) means time symmetry.

Arrow of Time: British astronomer Arthur Eddington coined the term *arrow of time* in 1927 to describe the one-way direction of time.

Ataraxia: "Suspending judgement led to a state of mind called 'ataraxia', quietude, peace of mind, or unperturbedness." (*Cambridge Dictionary of Philosophy*, p. 851). Skepticism is a practical philosophy concerned with leading us to a state of tranquility (ataraxia).

293

Autobiographical memory: Those who study the nature of memory notice two basic categories: memories that evoke a time and/or place, which has come to be called the *episodic memory* (also called the autobiographical memory), and all the other memories that are not stamped with a time or place are referred to as semantic memory.

Axial Age: Karl Popper called the period that included the Buddha, Lao Tse, Socrates, and Confucius the *Axial Age*. Recently, Karen Armstrong called this period *The Great Transformation* in a book by that title. Armstrong and others have expanded the period to be the 1st millennium BCE, but centered on the influence of these same people. This was a time when a greater sense of self-awareness came on the scene.

Bacon, Francis: Bacon (c.1561–c.1626) advocated the necessity of empirical observation and experimentation, which, combined with Cartesian doubt, laid the ground for modern science. He published *Novus Organum* in 1620.

Bhagavad Gita: (c. 5th Century BCE) perhaps the most influential text in the sacred literature of India. The sixth book of the *Mahabharata*.

Block Universe: The Block Universe is usually depicted as a sketch of a three-dimensional block, made up of an infinite number of time-slices (instances) that are arranged serially from the beginning to the end of time (think of a three-dimensional cube whose sides represent the boundary of all of space and time). The three-dimensional sketch shows two dimensions of space (x, y axes) while the third dimension (z axis) is made up of time-slices, representing all of time, *beginning* at the bottom and *ending* at the top.

Bodhidharma: (470–543?) Founder of Ch'an (Chinese Zen) school of Buddhism.

Cartesian: of or relating to René Descartes or his philosophy.

Childhood amnesia: Lack of memories of early childhood, generally before age four. Thought to be due to lack of sophisticated mental abilities, and in particular, language.

Citta (Chitta): In the Yogachara tradition of Buddhism the term is inclusive of the six consciousnesses (*vijñānas*), the sixth *vijñāna* is the sixth-sense of memory.

Cognitive sciences: The interdisciplinary scientific study of the mind that encompasses neuroscience, philosophy, linguistics, psychology, artificial intelligence, anthropology, and education.

Continental philosophy: philosophical traditions of the 19th and 20th century from mainland Europe. Mainly phenomenology, existentialism, hermeneutics, German idealism, post-structuralism, among others, but all are in general agreement that the natural sciences are *not* the only or most accurate way of understanding reality. Continental philosophy often stands in opposition to British empiricism.

Cybernetics: The science of communication and control theory. Concerning automatic control systems such as the nervous system and brain and mechanical-electrical systems.

Deconstruction: As the word suggests it means, a 'taking apart' or 'disassembling'. The term is most closely associated with Derrida's philosophy and with literary criticism. Originally, it is a term adapted from Heidegger's use of 'destruktion'. For Derrida, it is a linguistic maneuver that is employed to accomplish the deconstruction. However, the term has taken on broader implications. As used in *Time Sutra* it is the method of internal critique using the principles and concepts of a system to dismantle that same system, the system being, modernity.

Derrida, Jacques: (1930–2004) Founder of the controversial deconstructionist school of criticism. Associated with postmodernism and post-structuralism.

Descartes, René: (1596–1650) French natural philosopher and mathematician. Considered, along with Francis Bacon, the founder of modernity. Originated a method of skepticism (Cartesian Doubt) that would systematically uproot all prejudices and assumptions in an effort to discover the unshakable ground of truth. Devised the 'Cartesian coordinate system' that became a foundation of modern empirical science. His most important works on method: *Discourse on Method* (1637); *Meditations on First Philosophy* (1641); *Principles of Philosophy* (1644).

Dharma: In Buddhism the basic principles of existence, natural law, cosmic order. The teaching of the Buddha.

Dualism: A binary opposition. A theory that reality consists of two irreducible substances; Two opposing, mutually exclusive terms such as hot-cold and love-hate.

Duality: A dualism, a dichotomy. Two terms or ideas that stand in opposition to each other.

Dzogchen: Considered the ultimate teaching of the Nyingma tradition of Buddhism, Dzogchen is also known as Atiyoga. "Tib., lit., 'great perfection', the primary teaching of the Nyingmapa school of Tibetan Buddhism. This teaching, also known as ati-yoga (extraordinary yoga), is considered by its adherents as the definitive and most secret teaching of Shakyamuni Buddha ... According to the experience of *dzogchen* practitioners, purity of mind is always present and needs only to be recognized." (*Encyclopedia of Eastern Philosophy and Religion*, p. 97).

Emergence: The doctrine that properties at one level of organization are not predictable from the properties at a lower level of organization. As opposed to reductionism which holds that higher levels of organization can always be explained by the fundamental laws of physics, chemistry and biology.

Emergentist: One who accepts the doctrine of emergence.

Entropy: Entropy is the tendency of any ordered system to become increasingly disordered over time. The common usage of entropy has come to mean the tendency of everything to eventually degrade into disorder and chaos. The Second Law of Thermodynamics states that the entropy of an isolated system strives toward a maximum. Entropy is a mathematical factor that is the measure of the thermal energy that is unavailable to do work in a thermodynamic system.

Epiphenomenon: A secondary phenomenon accompanying and caused by a primary phenomenon, (i.e. the physical brain is the primary phenomenon, whereas *thought* is hypothesized to be a 'secondary', and thus epiphenomenon).

Episodic memory: Memories for individual episodes in someone's personal life. Those who study the nature of memory notice two basic categories; memories that evoke a time and or place, which has come to be called the *episodic memory* (also called the autobiographical memory), and all the other memories that are not stamped with a time or place and are referred to as semantic memory.

Eternal Recurrence (also Eternal Return): Nietzsche's theory of time taken from the Greek Stoic philosophers. The idea of recurring cyclic time in which all of history repeats endlessly.

Existentialism: "a philosophical and literary movement that came to prominence in Europe, particularly in France after World War II ... " (*Dictionary of the Philosophy of Science*, p. 296). This is the existential situation in which humans find themselves: thrown into an absurd world without meaning, nevertheless having to make a life out of it. Existentialism was dominantly a continental philosophy—the catastrophe of WWII, just one generation after the catastrophe of WWI, injected a sense of heightened urgency to find authentic existence in a meaningless world. "A chiefly 20th century philosophical movement embracing diverse doctrines but centering on analysis of individual existence in an unfathomable universe and the plight of the individual who must assume ultimate responsibility for his acts of free will without any certain knowledge of what is right or wrong or good or bad." (www.merriam-webster.com).

Explanatory Gap: The lack of an explanation of the *mental* in terms of the *physical*. The sciences (chemistry, physics, biology, etc.) to date have been lacking in their explanation of how mental phenomena are derived from a material world—this lack is the *Explanatory Gap*.

Great Transformation: Karl Popper called the period that included the Buddha, Lao Tse, Socrates, and Confucius the *Axial Age*. Recently, Karen Armstrong called this period *The Great Transformation* in a book by that title. Armstrong and others have expanded the period to be the 1st millennium BCE, but centered on the influence of these same people.

Heidegger, Martin: (1889–1976) German philosopher. His early work contributed to phenomenology and existentialism. His later work contributed to the development of hermeneutics. One of the most influential philosophers of the 20th century.

Hermeneutic circle: a closed system of self-referential parts (or symbols) that exhibit a circularity of interpretation such that every part gains its meaning from other parts within the system. This circle of interpretation cannot be escaped.

Isotropic time (isotropy): Time symmetry. Whereas, Anisotropy (of time) refers to a time in which past and future can be distinguished from each other.

Isotropy: Exhibiting properties with the same values when measured in different directions (along different axes). In reference to time the past and future cannot be distinguished from each other, thus time is symmetric (isotropic).

Koan: In Zen, a paradox to be meditated upon, and when resolved will lead to insight.

Lao Tzu: (6th century BCE) Chinese philosopher; considered founder of Taoism and author of *Tao-te-Ching*.

Longchenpa: (1308–1364) Tibetan Buddhist philosopher; indisputably the greatest scholar of the Nyingma tradition. His work brought a systematic and philosophical organization to the 600 years of Nyingma literature that preceded him, and culminates in the exposition of Dzogchen.

Long-term memory: Permanent memory. Semantic memory for facts along with episodic (autobiographical) memory.

Manovijñāna: Sanskrit term for the mental *sixth consciousness* (the sixth sense).

Materialist: One who accepts the thesis that whatever exists, is, or depends upon, matter. Materialists maintain that the entire mind-body problem is ultimately reducible to physics, chemistry, and biology.

Meme: (from Richard Dawkins, 1976). "Skills, habits, songs, stories, or any other kind of information that is copied from person to person." (*Oxford Companion To The Mind*, p. 558). A meme (rhymes with *seem*) acts much like a gene in that it contains a unit of information, but rather than information residing in DNA or RNA, the meme resides in human memory, and is passed from one human brain to another.

Memetics: The science that is the study of memes.

Memory is past: A phrase used throughout *Time Sutra* as a reminder that the Representative Theory of Memory is an assumption. A mnemonic device to bring to mind the fact that memory is always an experience of the *present*.

Memory Theory of Identity: The idea that our memories are identical with what constitutes our sense of self. This was John Locke's 1690 view. We are what we remember ourselves to be. The self is constructed out of memory.

Mnemonic device: Something intended to assist memory. An aid in calling to mind.

Modernity: Broadly, the post-medieval historical period. Characterized by clocks, capitalism, industrialization, secularization, rationalism, and the nation state. For the purposes of *Time Sutra*, the nearly four centuries since the advent of Cartesian philosophy.

Monism: The view that there is only one kind of ultimate substance. The view that all of reality is one unified organic whole; The philosophy or theory that reduces all phenomena to one principle.

Nagarjuna: 2nd/3rd century Indian. One of the most important philosophers of Buddhism and the founder of the Madhyamika school of Buddhism.

Nietzsche, Friedrich: (1844–1900) German philosopher. His philosophy became very influential in the 20th century. His philosophy of time is discussed in some detail in chapter 15.

Nihilism: a doctrine that denies any objective ground of truth. The view that traditional beliefs are unfounded.

Nirvana: a state of liberation or illumination. The goal of spiritual practice in all branches of Buddhism. This entails realization of the true nature of mind along with the elimination of ego.

Parmenides of Elea: (born c. 515 BCE) Greek philosopher, the most influential of the pre-Socratic philosophers and exerted the deepest influence on Plato's thought. The first systematic philosopher of the Western Tradition whose thought is known from a single poem titled *On Nature*.

Parmenides Maxim: 'What *is*, is, and cannot not be', or said another way, 'only being—what is—can exist'. Parmenides is declaring 'Being is', and non-being is impossible, and this makes reality a timeless, changeless, all-in-One.

Past Hypothesis: The Big Bang Theory of the cosmos assumes a very low entropy of the early universe. David Albert (2000) has given this assumption a simple name: the *Past Hypothesis*.

Perennial Philosophy: A term first used by Agostino Steuco (1497–1548). Leibniz brought the term into prominence. This refers to a form of philosophy that has recurred for over two thousand years in both East and West. The perennial philosophy also argues that all theologies share a set of common fundamental attributes.

Piaget, Jean: (1896–1980) Swiss psychologist. Leading figure in the field of child psychological development. In particular he described how the developing child learns the concepts of time and space.

Reduction, reductionism: In philosophy of science, reduction is the explanation of a more complex phenomenon in terms of more fundamental, basic or general ones. In biology, the belief living phenomena can be understood in terms of the laws of physics and chemistry.

Reductionist: of or relating to the theory of reductionism.

Representative Theory of Memory: The theory that our memory, although always a present experience, is representative of an experience in the past. In *Time Sutra* the theory is represented by the meme '*memory is past.*'

Samsara: Sanskrit term used in the East to mean the illusory world in which humans are caught up. The cycle of existence of birth and death. The world of suffering.

Satori: a state of intuitive illumination sought in Zen Buddhism.

Second Law of Thermodynamics (also 2nd Law): States that the entropy of an isolated system strives toward a maximum. Entropy, over time, always increases.

Semantic memory: "long-term memory for facts, other than autobiographical." (*Oxford Companion to the Mind*, p. xviii).

Short-term memory: Working memory, immediate memory, primary memory.

Sixth sense: In general this is the psychic or mental sense as opposed to the five bodily senses (sight, smell, taste, touch, hearing). Often referred to as *mind*, but in the context in which it is being used in *Time Sutra* is specifically the faculty of *memory*. (see chapter 8) . *Time Sutra* is declaring that the schism of the senses is responsible for the mind-body problem, and further, to heal this schism requires the elimination of time and ego.

Smriti (smṛti): Sanskrit term for *memory,* but the intended Buddhist meaning of *smriti* is most certainly far different from the Western concept of memory. Generally it means mindfulness, attention, and recollection.

Solipsism: a theory that the self can know nothing but its own experience and that the self is the only existent thing.

Spatiotemporal: Having both spatial and temporal qualities. Of or relating to spacetime.

Substance duality: the theory that reality is composed of two different and seemingly incommensurable substances—*mind* and *matter*.

Sutra: Sanskrit term that literally means thread. In Eastern philosophy the *sutra* was used as a mnemonic device that would prompt a remembrance of the 'thread of the argument' that a particular, often lengthy, religious text was trying to convey. In the context of *Time Sutra* it is a thread of reason that divulges the nature of time.

Temporal: of or relating to time. That which is in time.

Temporality: The quality or state of being temporal.

Triple Time: Sanskrit term (*trikala*) can simply mean the mundane 'world of the three times' of the past, present and future.

Veda (Sanskrit): "knowledge". The term used to denote the large body of sacred literature of Aryan Indians.

Vedanta: an orthodox system of Hindu philosophy.

Vedic: of or relating to the Vedas.

Vijñāna: Sanskrit term for *consciousness*, of which there are six; the five sensory and one mental consciousness (the *Manovijñāna* or sixth sense).

Vitalism: A doctrine stating that the functions of a living organism are due to a vital principle distinct from physicochemical processes.

Will to Power: A central tenet of Nietzsche's philosophy which is closely tied in with his concept of time (see discussion in chapter 15).

Working memory: Short-term memory, immediate memory, primary memory.

Zazen: Japanese, literally 'sitting-zen'.

Zombie, philosophical (two types): a philosophical zombie is an android that behaves like a human but is not conscious. *Time Sutra* poses a second type of philosophical zombie: an android that behaves like a human and is conscious.

BIBLIOGRAPHY

Works referred to in the text.

Arendt, Hannah. 1997. *The Life of the Mind*, vols. 1 & 2. Harcourt.

Armstrong, Karen. 2006. *The Great Transformation: The Beginning of our Religious Traditions*. Knopf.

Atmanspacher, Harald and Eva Ruhnau, eds. 1997. *Time, Temporality, Now: Experiencing Time and Concepts of Time in an Interdisciplinary Perspective*. Springer-Verlag.

Audi, Robert, general ed. 1999. *Cambridge Dictionary of Philosophy*, 2nd ed. Cambridge University Press.

Augustine, Saint, Bishop of Hippo. F. J. Sheed, trans. 1942. *Augustine: Confessions, Books I-XIII*. Hackett Publishing Co.

Aveni, Anthony F. 1995. *Empires of Time: Calendars, Clocks, and Culture*. Kodansha International.

Ayer, A. J. 1952. *Language, Truth and Logic*. Dover Publications.

Balsekar, Ramesh S. 1992. *Consciousness Speaks*. Advaita Press.

Balslev, Anandita Niyogi. 1999. *A Study of Time in Indian Philosophy*. Munshiram Manoharlol Press.

Barbour, Julian. 1999. *The End Of Time: The Next Revolution in Physics*. Oxford University Press.

Bareau, André. 1991. in H. S. Prasad, ed. *Essays on Time in Buddhism*. Sri Satguru Publications.

Batchelor, Stephan. 2002. in Alan Hunt Badiner, ed. *Zig Zag Zen: Buddhism and Psychedelics*. Chronicle Books.

Beckett, Samuel. 1965. *Proust and Three Dialogues with Georges Duthuit*. John Calder Publishers.

Beiser, Frederick C. 1987. *The Fate of Reason: German Philosophy from Kant to Fichte*. Harvard University Press.

Bergson, Henri. F. L. Pogson, trans. 1910. *Time and Free Will*. The Macmillan Co.

Berman, Morris. 1990. *Coming To Our Senses: Body and Spirit in the Hidden History of the West*. Bantam Books.

Blackmore, Susan. 1999. *The Meme Machine*. Oxford University Press.

Block, Ned. 1996. in Donald M. Borchert, ed. *The Encyclopedia of Philosophy, Supplement*. Simon & Schuster Macmillan.

Blofeld, John, trans. 1958. *The Zen Teachings of Huang Po*. Grove Press.

Bloom, Harold. 2004. *Where Shall Wisdom Be Found?* Riverhead Books.

Bodhidharma. Red Pine. trans. 1989. *The Zen Teachings of Bodhidharma*, North Point Press.

Borchert, Donald M., ed. 1996. *The Encyclopedia of Philosophy, Supplement.* Simon & Schuster Macmillan.

Borges, Jorges Luis. 1986. "A New Refutation of Time." In *Labyrinths: Selected Stories and Other Writings.* New Directions Publishing.

———. 1998. "Book of Sand." In *Collected Fictions.* Penguin.

———. 2000. *The Library of Babel.* David R. Godine Publisher.

Bradley, F. H. 1893. *Appearance and Reality.* Oxford University Press.

Brentano, Franz. 1988. Barry Smith, trans. *Philosophical Investigations in Space, Time and the Continuum.* Croom Helm.

Brodie, Richard. 1996. *Virus of the Mind: The New Science of the Meme.* Hay House.

Buswell, Robert E. Jr. And Donald S. Lopez Jr. 2014. *The Princeton Dictionary of Buddhism.* Princeton University Press.

Bynum, William F., ed. 1989. *Dictionary of the History of Science.* Princeton University Press.

Callender, Craig. 2010. "Is Time an Illusion?" *Scientific American*, June 2010.

Campbell, Joseph. 1968. *Creative Mythology: The Masks of God.* Penguin.

Camus, Albert. 1970. Ellen Conroy Kennedy, trans. *Lyrical and Critical Essays.* Vintage Books.

Čapek, Milič. 1986. in David Ray Griffin ed. *Physics and the Ultimate Significance of Time.* State University of New York.

Capra, Fritjof. 1997. *Web of Life: A New Scientific Understanding of Living Systems*. Anchor Books.

Carroll, Sean. 2010. *From Eternity to Here: The Quest for the Ultimate Theory of Time*. Dutton.

Carse, James P. 1986. *Perfect Brilliant Stillness: Beyond the Individual Self*. Paragate Publishing.

———. 1994. *Breakfast at the Victory: The Mysticism of the Ordinary Experience*. Harper One.

Casey, Edward S. 1987. *Remembering : A Phenomenological Study*. Indiana University Press.

Chalmers, David J. 1996. *The Conscious Mind: In Search of a Fundamental Theory*. Oxford University Press.

Cleary, Thomas, trans. 1978. *Original Face: An Anthology of Rinzai Zen*. Grove Press Inc.

———. 1993. *No Barrier: The Munmonkan*. Bantam Books.

———. 1997. *Kensho: Several Works by Haukuin*. Shambala.

Cleugh, M. F. 1937. *Time: and Its Importance in Modern Thought*. Methuen & Co. LTD. London.

Collingwood, R. G. 1972. in Paul Edwards, ed. *The Encyclopedia Of Philosophy, vol. 2*. Macmillan Publishing.

Coleman, Graham, ed. 1994. *A Handbook of Tibetan Culture: A Guide to Tibetan Centres and Resources Throughout the World*. Shambala.

308

Coleman, James William. 2001. *The New Buddhism: The Western Transformation of an Ancient Tradition.* Oxford University Press.

Copleston, Frederick. 1993. *A History of Philosophy*, *vol.1 Greece and Rome.* Image Books.

Coveney, Peter and Roger Highfield. 1990. *The Arrow of Time.* Ballantine Books.

Crane, George. 2005. *Beyond the House of the False Lama.* Harper Collins Publishers.

Csikszentmihalyi, Mihaly. 1990. *Flow: The Psychology of Optimal Experience.* Harper Perennial.

Dalai Lama, 14th. 2000. *Dzogchen: The Heart Essence of Great Perfection.* Snow Lion Publications.

Darling, David J. 2004. *The Universal Book of Mathematics: From Abracadabra to Zeno's Paradoxes.* John Wiley and Sons.

David-Neel, Alexandra and Lama Yongden. 1967. *The Secret Oral Teachings In Tibetan Buddhist Sects.* City Lights Books.

Davies, Paul. 1995. *About Time.* Simon and Schuster.

Dawkins, Richard. 1976. *The Selfish Gene.* Oxford University Press.

De Zengotita, Thomas. 2003. Common Ground: Finding Our Way Back to the Enlightenment, *Harper's*, Jan. 2003

Dennett, Daniel, 1998. *Brainchildren*: *Essays on Designing Minds.* MIT. Press.

——. 2005. *Sweet Dreams: Philosophical Obstacles to a Science of Consciousness.* MIT Press.

——. 2007. *Breaking the Spell: Religion as a Natural Phenomenon.* Penguin Books.

Descartes, René. 1986. John Cottingham, ed. and trans. *Meditations on First Philosophy.* Cambridge University Press.

Detienne, Marcel. Janet Lloyd, trans. 1999. *The Masters of Truth in Archaic Greece.* Zone Books.

Dogen Zenji. Kosen Nishiyama, trans. 1975. *Shobogenzo, vol.1.* Nakayama Shobo.

——. 2004. Leighton, Taigen Dan and Shohaku Okumura trans. *Dogen's Extensive Record: A Translation of the Eihei Koroku.* Wisdom Publications.

Dokic, Jerome. 2001. in Hoerl and McCormack, eds. *Time And Memory: Issues in Philosophy and Psychology.* Oxford University Press.

Dossey, Larry. 1982. *Time, Space, and Medicine.* Shambala.

Dowden, Bradley, in *The Internet Encyclopedia of Philosophy.* Originally published August 11, 2001. <http://www.iep.utm.edu/time/>

Durant, Will. 1926. *The Story of Philosophy.* Simon and Schuster.

Edelman, Gerald. 2004. *Wider Than The Sky: The Phenomenal Gift of Consciousness.* Yale University Press.

Edmonds, David and John Eidinow. 2001. *Wittgenstein's Poker.* Harper Collins.

Edwards, Paul, ed. in chief. 1972. *The Encyclopedia of Philosophy*. Macmillan Publishing.

Einstein, Albert. 1982. *Ideas and Opinions*. Crown Publishing Group.

Einstein, Albert and Leopold Enfield. 1938. *The Evolution of Physics: From Early Concepts to Relativity and Quanta*. Simon and Schuster.

Eliade, Mircea. 1952 , *Images and Symbols*. Princeton University Press.

———. 1954, *The Myth of the Eternal Return*. Princeton University Press.

———. 1957. *The Sacred and the Profane*. Harcourt.

———. 1960. *Myth, Dreams, and Mysteries*. Harper and Row.

———. 1963. *Myth and Reality*. Waveland Press.

———. 1992. In H. S. Prasad, ed. *Time in Indian Philosophy*: *A Collection of Essays*. Sri Satguru Publications.

Eliot, T. S. 1943. *Four Quartets*. Harcourt, Brace and World Inc.

———. 1964. *Knowledge and Experience in the Philosophy of F. H. Bradley*. Columbia University Press.

Empiricus, Sextus. 2005. Richard Bett, trans. and ed. *Against The Logicians*. Cambridge University Press.

Evans-Wentz, W. Y. 2000. *The Tibetan Book of the Great Liberation*. Oxford University Press.

Falk, Dan. 2008. *In Search of Time: The Science of a Curious Dimension.* St. Martins Press.

Feyerabend, Paul. 1993. *Against Method.* Verso.

Feynman, Richard P. 1998. *The Meaning of It All: Thoughts of a Citizen Scientist.* Helix Books.

Figal, Gunter. In Robert J. Dostal, ed. 2002. *The Cambridge Companion to Gadamer.* Cambridge University Press.

Fraser, J. T., ed. 1966. *The Voices of Time.* George Braziller, Inc.

Fraser, J. T., et al., eds. 1972. *The Study of Time.* Springer-Verlag.

————. 1978. *The Study of Time III.* Springer-Verlag.

————. 1988. *Time: the Familiar Stranger.* Tempus Books.

Friedman, William J. 1990. *About Time: Inventing the Fourth Dimension.* MIT Press, 1990.

Fulbrook, Mary. 2002. *Historical Theory.* Routledge.

Gallagher, Shaun and Jonathon Shear, eds. 1999. *Models of the Self.* Imprint Academic.

Gallop, David, trans. 1984. *Parmenides of Elea.* University of Toronto Press.

Geldard, Richard. 2000. *Remembering Heraclitus.* Lindisfarne Books.

————. 2007. *Parmenides and The Way of Truth.* Monkfish Publishing.

Bibliography

Gell, Alfred. 1992. *The Anthropology of Time: Cultural Constructions of Temporal Maps and Images.* Berg.

Genoud, Charles. 2006. *Gesture of Awareness: A Radical Approach to Time, Space, and Movement.* Wisdom Publications.

Giddens, Anthony. 1990. *The Consequences of Modernity.* Stanford University Press.

Goldstein, Joseph. *Tricycle: The Buddhist Review.* Winter 2001.

Gray, Jeffrey. 2004. *Consciousness: Creeping Up on the Hard Problem.* Oxford University Press.

Greene, Brian. 2004. *The Fabric of the Cosmos: Space, Time, and the Texture of Reality.* Alfred A. Knopf.

Gregory, Richard L., ed. 2004. *The Oxford Companion to the Mind.* Oxford University Press.

Griffin, David R., ed. 1986. *Physics and the Ultimate Significance of Time.* State University of New York Press.

———. 1988. *The Reenchantment of Science: Postmodern Proposals.* State University of New York Press.

Griffiths, Jay. 2004. *A Sideways Look At Time.* Penguin Books.

Grosz, Elizabeth. 1999. *Becomings: Explorations in Time, Memories, and Futures.* Cornell University Press.

Gunaratana, Henapola. 1992. *Mindfulness: in Plain English.* Wisdom Publications.

313

Guthrie, W. K. C. 1960. *The Greek Philosophers: from Thales to Aristotle.* Harper Torchbooks.

———. 1965. *A History of Greek Philosophy: vol. II.* Cambridge University Press.

Hamilton, Edith and Huntington Cairns, eds.. 1961, *The Collected Dialogues of Plato.* Pantheon Books.

Harris, Errol E. 1988. *The Reality of Time.* State University of New York.

Hassard, John. 1994. in Samuel L. Macey, ed. *Encyclopedia of Time.* Garland Publishing Inc.

Hawking, Stephen. 1988. *A Brief History of Time: from the Big Bang to Black Holes.* Bantam Books.

Haxton, Brooks, trans. 2001. *The Collected Wisdom of Heraclitus.* Penguin Books.

Heidegger, Martin. 1962. John Macquarrie and Edward Robinson, trans. *Being and Time.* Harper and Row Publishers.

———. 1976. J. Glen Gray, trans. *What Is Called Thinking.* Perennial Library.

———. 1998. André Schuwer and Richard Rojcewicz, trans. *Parmenides.* Indiana University Press.

———. 2002. Ted Sadler, trans. *The Essence of Truth.* Continuum.

Heidegger, Martin and Eugen Fink. 1993. Charles H. Seibert, trans. *Heraclitus Seminar.* Northwestern University Press.

Heine, Steven. 1985. *Existential and Ontological Dimensions of Time in Heidegger and Dogen.* State University of New York Press.

Hoerl, Christoph and Teresa McCormack, eds. 2001. *Time and Memory: Issues in Philosophy and Psychology.* Oxford University Press.

Hofstadter, Douglas. 1979. *Gödel, Escher, Bach: An Eternal Golden Braid.* Basic Books.

———. 1999. *Gödel, Escher, Bach: An Eternal Golden Braid*: 20th Anniversary Edition. Basic Books.

———. 2007. *I Am a Strange Loop.* Basic Books.

Horgan, John. 2000. *The Undiscovered Mind: How the Human Mind Defies Replication, Medication, and Explanation.* Simon and Schuster.

Horkheimer, Max and Theodor W. Adorno. 2002. John Cumming, trans. *Dialectic of Enlightenment.* Continuum.

Hull, David. 1988. in B. Weber, D. Depew and J. Smith, eds. *Entropy, Information, and Evolution.* MIT Press.

Huntington, C. W. with Geshe Namgyal Wangchen. 1989. *The Emptiness of Emptiness: An Introduction to Early Madhyamika.* University of Hawaii Press.

Husserl, Edmund. 1991. John Barnett Brough, trans. *On the Phenomenology of the Consciousness of Internal Time (1893–1917).* Kluwer Academic Publishers.

Huxley, Aldous. 1945. *The Perennial Philosophy.* Harper and Brothers Publishers.

Jackson, Kevin. 2007. *The Book Of Hours*. Duckworth Overlook.

Jaini, Padmanabh S. 1992. In Janet Gyatso, ed. *In The Mirror Of Memory: Reflections on Mindfulness and Remembrance in Indian and Tibetan Buddhism*. State University of New York Press.

James, William. 1971. *Essays in Radical Empiricism: And a Pluralistic Universe*. Dutton.

———. 1987. "Varieties of Religious Experience" in *William James: Writings, 1902–1910*. Library of America.

Jaspers, Karl. 1957. Hannah Arendt, ed., Ralph Manheim, trans. *Plato and Augustine–from the Great Philosophers, vol. 2*. Harvest Books.

———. 1966. Hannah Arendt, ed., Ralph Manheim, trans. *Anaxamander, Heraclitus, Parmenides, Plotinus, Lao-Tzu, Nagarjuna*. From *The Great Philosophers, vol. 2*. Harvest Books.

———. 1997. C. F. Wallraff and F. J. Schmitz, trans. *Nietzsche: an Introduction to an Understanding of his Philosophical Activity*. Johns Hopkins Press.

Jaynes, Julian. 1976. *The Origin of Consciousness in the Breakdown of the Bicameral Mind*. Houghton Mifflin Company.

Kahn, Charles H. 2009. *Essays on Being*. Oxford University Press.

Katagiri, Dainin. 2007. *Each Moment is the Universe: Zen and the Way of Being Time*. Shambala.

Kearney, Richard, ed. 1997. *Routledge History of Philosophy, vol. 8*. Routledge.

Keizan. 1990. Thomas Cleary, trans. *Transmission of Light: Zen in the Art of Enlightenment.* North Point Press.

Kenny, Anthony. 2010. *A New History of Western Philosophy.* Oxford University Press.

Klein, Stefan. 2006. *The Secret Pulse of Time: Making Sense of Life's Scarcest Commodity.* Marlowe and Company.

Kotre, John. 1996. *White Gloves: How We Create Ourselves Through Memory.* Norton and Company.

Kriegel, Uriah and Kenneth Williford. 2006. *Self-Representational Approaches to Consciousness.* MIT Press.

Krishnamurti, J. 1969. *Freedom From The Known.* Harper Collins.

———. 1975. *The First And Last Freedom.* Harper and Row.

Kuhn, Thomas S. 1962. *The Structure of Scientific Revolutions.* University of Chicago Press.

Leach, E. R. 1961. *Rethinking Anthropology.* The Athlone Press.

LeDoux, Joseph. 2002. *Synaptic Self: How Brains Become Who We Are.* Viking Press.

Libet, Benjamin. 2004. *Mind Time: The Temporal Factor in Consciousness.* Harvard University Press.

Lloyd, Genevieve. 1993. *Being in Time: Selves and Narrative in Philosophy and Literature.* Routledge.

Longchenpa. 1998. Richard Barron, trans. *Way Of Abiding*. Padma Press.

———. 2001. Richard Barron, trans. *A Treasure Trove of Scriptural Transmission: A Commentary on The Basic Space of Phenomena*. Padma Publishing.

———. 2010. Keith Dowman, trans. *Natural Perfection: Longchenpa's Radical Dozgchen*. Wisdom Publications.

———. 2010. Kennard Lipman and Merrill Peterson, trans. *You Are the Eyes of the World*. Snow Lion Publications.

Lopez, Donald. 2002. *Tricycle: The Buddhist Review*. Fall.

Lovejoy, Arthur O. 1955. *The Revolt Against Dualism*. Open Court.

Loy, David. 1986. 'The Mahayana Deconstruction of Time' in *Philosophy East and West*, Jan. 1986.

———. 1988. *Nonduality*. Humanity Books.

———. 1996. *Lack and Transcendence: The Problem of Death and Life in Psychotherapy, Existentialism, and Buddhism*. Humanity Books.

———. 2002. *A Buddhist History of the West: Studies in Lack*. State University of New York Press.

———. 2009. *Awareness Bound and Unbound: Buddhist Essays*. State University of New York Press.

Lucas, J. R. 1999. 'A Century of Time' in Jeremy Butterfield, ed. *The Arguments of Time*. Oxford University Press.

Lucretius, Titus Carus. 1968. Rolfe Humphries, trans. *Lucretius: The Way Things Are.* Indiana University Press.

Lyotard, Jean-François. 1991. Bennington and Bowlby, trans. *The Inhuman.* Stanford University Press.

Macey, Samuel L. 1991. *Time: A Bibliographic Guide.* Garland Publishing.

Macey, Samuel L., ed. 1994. *Encyclopedia of Time.* Garland Publishing.

Magnus, Bernd and Kathleen M. Higgins, eds. 1996. *The Cambridge Companion to Nietzsche.* Cambridge University Press.

Mahadevan, T. M. P. 1992. in H. S. Prasad, ed. *Time in Indian Philosophy: A Collection of Essays.* Sri Satguru Publications.

Mann, Thomas. 1939. *Schopenhauer.* Longmans, Green and Company.

Martin, M. G. F. 2001. in Hoerl and McCormack, eds. *Time and Memory: Issues in Philosophy and Psychology.* Oxford University Press.

Mascaro, Juan, trans. 1962. *Bhagavad Gita.* Penguin Books.

Maugham, W. Somerset. 1984. *The Razor's Edge.* Penguin Books.

McDermott, John J. 1994. in Samuel L. Macey, ed. *Encyclopedia of Time.* Garland Publishing.

McIntosh, Anthony R. 2007. in Henry L. Roediger, Dudai and Fitzpatrick, eds. *Science of Memory: Concepts.* Oxford University Press.

McKeon, Richard. 1975. in Charles M. Sherover, ed. *The Human Experience of Time: The Development of Its Philosophic Meaning.* Northwestern University Press.

McLeod, Melvin, ed. 2011. *Best Buddhist Writing 2011.* Shambala.

McMahan, David L. 2008. *The Making of Buddhist Modernism.* Oxford University Press.

McTaggart, John. 1927. *The Nature of Existence, vol.2.* Cambridge University Press.

Mehlberg, Henry. 1980. *Time, Causality, and the Quantum Theory, vol. 1.* D. Reidel Publishing Company.

———. 1980. *Time, Causality, and the Quantum Theory, vol. 2.* D. Reidel Publishing Company.

Mlodinow, Leonard. 2012. *Subliminal: How Your Unconscious Mind Rules Your Behavior.* Vintage Books.

Montaigne, Michel de. 1987. M. A. Screech, trans. *An Apology for Raymond Sebond.* Penguin Books.

Morris, Richard. 1986. *Time's Arrow: Scientific Attitudes Toward Time.* Touchstone Books.

Mourelatos, Alexander P. D. 1974. *The Route of Parmenides.* Yale University Press.

Mumford, Lewis. 1934. *Technics and Civilization.* University of Chicago Press edition 2010.

———. 1983. in David S. Landes. *Revolution in Time*: *Clocks and the Making of the Modern World.* Harvard University Press.

Mundle, C.W.K. 1972. in Paul Edwards, ed. *The Encyclopedia Of Philosophy, vol. 8.* Macmillan and Company.

Nagarjuna. 1987. Venkata K. Ramanan, trans. "Nagajuna's Philosophy." as presented in *The Maha-Prajnaparamita-Sastra.* Motilal Banarsidass.

———. 1987. David Ross Komito, trans. *Nagarjuna's Seventy Stanzas: A Buddhist Psychology of Emptiness.* Snow Lion Publications.

Nagarjuna and Kaysang Gyatso. 1975. Jeffrey Hopkins, et al., trans. *The Precious Garland and The Song of the Four Mindfulnesses.* Harper and Row Publishers.

Nakamura, Hajime. 1966. 'Time in Indian and Japanese Thought' in J. T. Fraser, ed. *The Voices of Time.* George Braziller.

Needleman, Jacob. 2003. *Time And The Soul.* Berrett-Koehler

Neisser, Ulric and Ira E. Hyman, Jr. 2000. *Memory Observed: Remembering in Natural Contexts.* Worth Publishers.

Nelson, Katherine. 2001. in Daniel L. Schacter and Elain Scarry, eds. *Memory, Brain, and Belief.* Harvard University Press.

Nietzsche, Friedrich. 1990. R. J. Hollingdale, trans. *Twilight of the Idols.* Penguin Books.

———. 1997. R. J. Hollingdale, trans. *Untimely Meditations.* Cambridge University Press.

Nicoll, Maurice. 1953. *Living Time.* Hermitage House.

Nisargadatta Maharaj. 1973. *I Am That.* Acorn Press.

Nixon, Gregory. 1999. Francisco Varela and Jonathan Shear, eds. *The View From Within: First Person Approaches to the Study of Consciousness.* Imprint Academic.

Noll, Richard and Carol Turkington. 1994. *Encyclopedia of Memory and Memory Disorders.* Facts On File.

Norbu, Namkhai. 2000. *The Crystal and the Way of Light: Sutra, Tantra, and Dzogchen.* Snow Lion.

Nyima, Chokyi, Rinpoche. 1989. Erik Hein Schmidt, trans. *The Union of Mahamudra and Dzogchen.* Snow Lion.

Oelschlaeger, Max. 1991. *The Idea Of Wilderness.* Yale University Press.

Ornstein, Robert E. 1997. *On The Experience Of Time.* Westview Press.

Owen, G. E. L. 1974. "Plato and Parmenides on the Timeless Present." In Alexander P. D. Mourelatos, ed. *The Pre-Socratics: A Collection of Critical Essays.* Odyssey Press.

Padmasambhava. 1990. Erik Pema Kunsang, trans. *Dakini Teachings*: *Padmasambhava's Oral Instructions to Lady Tsogyal.* Shambala.

Pagels, Elaine. 2003. *Beyond Belief: The Secret Gospel of Thomas.* Random House.

Panikkar, Raimundo. 1978. in J. T. Fraser, et al., eds. *The Study of Time III.* Springer-Verlag.

———. 1992. In H. S. Prasad, ed. *Time in Indian Philosophy*: *A Collection of Essays*. Sri Satguru Publications.

Panksepp, Jaak, 1999. in Shaun Gallagher and Jonathon Shear, eds. *Models of the Self*. Imprint Academic.

Park, David. 1972. "The Myth of the Passage of Time.*"* In J. T. Fraser, et al., eds. *The Study of Time, vol.1*. Springer-Verlag.

———. 1980. *The Image Of Eternity: Roots of Time in the Physical World*. University of Massachusetts Press.

Penrose, Roger. 1989. *The Emperor's New Mind*. Oxford University Press.

Piaget, Jean. 1967. F. J. Langdon and L. L. Lunzer, trans. *The Child's Conception of Space*. W. W. Norton and Company.

———. 1969. A. J. Pomerons, trans. *The Child's Conception of Time*. Ballantine Books.

Plato. 1961. Edith Hamilton, and Huntington Cairns, eds. *The Collected Dialogues of Plato*. Pantheon Books.

Ponlop, Dzogchen. 2006. Melvin McLeod, ed. *Best Buddhist Writing 2006.* Shambala.

Popper, Karl. 1966. *The Open Society and Its Enemies*. vols.1 & 2. Princeton University Press.

———. 1985. *Unended Quest*. Open Court.

———. 2002. *The World of Parmenides*. Routledge.

Prasad, H. S. ed. 1991. *Essays on Time in Buddhism.* Sri Satguru Publications.

———. ed. 1992. *Time in Indian Philosophy: A Collection of Essays.* Sri Satguru Publications.

Price, Huw. 1990. Review of the Arrow of Time and Time's Journeys, in *Nature* 348. Nov., p. 356, 22, November.

———. 1996. *Time's Arrow and Archimedes Point: New Directions for the Physics of Time.* Oxford University Press.

Priestly, J. B. 1968. *Man and Time.* Dell Publishing Company Inc.

Radday, Yehuda T. and Heim Shore. 1985. *Genesis: An Authorship Study in Computer-Assisted Statistical Linguistics.* Biblical Institute Press.

Radhakrishnan, Sarvepalli and Charles A. Moore, eds. 1957. *A Sourcebook in Indian Philosophy.* Princeton University Press.

Rangdrol, Tsele Natsok. 1989. Erik Pema Kunsang, trans. *Lamp of Mahamudra.* Shambala.

Reyna, Ruth. 1992. In H. S. Prasad, ed. *Time in Indian Philosophy: A Collection of Essays.* Sri Satguru Publications.

Ruhnau, Eva. 1997. "The Deconstruction of Time and the Emergence of Temporality." in Harald Atmanspacher et al., eds., *Time, Temporality, Now: Experiencing Time and Concepts of Time in an Interdisciplinary Perspective.* Springer-Verlag.

Russell, Bertrand. 1921. *The Analysis of Mind.* IndyPublish.com

———. 1945. *A History of Western Philosophy.* Simon and Schuster.

———. 1954 *The Analysis of Matter*. Dover Publications.

———. 1959. *My Philosophical Development*. Simon and Schuster.

Ryokan. 1996. Ryuichi Abe and Peter Haskell, trans. *Great Fool: Zen Master Ryokan: Poems, Letters, and Other Writing*. University of Hawaii Press.

Sachs, Robert G. 1987. *The Physics of Time Reversal*. University of Chicago Press.

Sanches, Francisco. 1988. *That Nothing Is Known*. Cambridge University Press.

Schacter, Daniel L. 1996. *Searching for Memory: The Brain, Mind, and the Past*. Basic Books.

Schrödinger, Erwin. 1954. *Nature and the Greeks: and Science and Humanism*. Cambridge University Press.

Schumacher, Stephan and Gert Woerner, eds. 1989. *Encyclopedia of Eastern Philosophy and Religion*. Shambala.

Scruton, Roger. 1995. *A Short History of Modern Philosophy: From Descartes to Wittgenstein*. Routledge

———. 1999. *Spinoza*. Routledge.

Seigel, Jerrold. 2005. *The Idea of the Self: Thought and Experience in Western Europe Since the Seventeenth Century*. Cambridge University Press.

Sherover, Charles M., ed. 1975. *The Human Experience of Time: The Development of Its Philosophical Meaning*. Northwestern University.

Silverman, Hugh J. 1997. Richard Kearney, ed. in *Routledge History of Philosophy*, vol. 8. Routledge.

Sinha, Braj M. 1983. *Time and Temporality in Samkhya-Yoga and Abhidharma Buddhism*. Munshiram Manoharlal Pub.

Sklar, Lawrence. 1985. in, *Philosophy and Spacetime Physics*. University of California Press.

———. 1993. in Robin Le Poidevin, and Murray MacBeath, eds. *The Philosophy of Time*. Oxford University Press.

Snyder, Gary. 1995, in Carole Tonkinson, ed. *Big Sky Mind: Buddhism and the Beat Generation*. Riverhead Books.

Soeng, Mu. 2004. *Trust in Mind: The Rebellion of Chinese Zen*. Wisdom Publications.

Sokowolski, Robert. 1999. in Robert Audi, General ed. *Cambridge Dictionary of Philosophy*, 2nd ed. Cambridge University Press.

Sontag, Susan. 1961. *Against Interpretation and Other Essays*. Picador.

Soseki, Muso. 1989. *Sun At Midnight*. Northpoint Press.

Stambaugh, Joan. 1987. *The Problem of Time in Nietzsche*. Associated University Presses.

———. 1999. *The Formless Self*. State University of New York Press.

Stenger, Victor J. 2000. *Timeless Reality: Symmetry, Simplicity, and Multiple Universes*. Prometheus.

Suzuki, D. T. 2000. *Outlines of Mahayana Buddhism*. Munshiram Manoharlal Publishing.

Taylor, C. C. W. ed. 1997. *Routledge History of Philosophy, vol.1: From the Beginning to Plato*. Routledge.

Taylor, Charles. 1989. *Sources of the Self: The Making of the Modern Identity*. Harvard University Press.

Thanassas, Panagiotis. 2007. *Parmenides, Cosmos, and Being: A Philosophical Interpretation.* Marquette University Press.

Tolle, Eckhart. 1999. *The Power of Now: A Guide to Spiritual Enlightenment*. New World Library.

Torey, Zoltan. 2009. *The Crucible of Consciousness: An Integrated Theory of Mind and Brain.* The MIT Press.

Trivers, Howard. 1985. *The Rhythm of Being: A Study of Temporality.* Philosophical Library.

Trungpa, Chogyam. 1988. *The Myth of Freedom: and The Way of Meditation.* Shambala.

———. 2001. *Glimpses of the Abhidharma.* Shambala.

Turetzky, Phillip. 1998. *Time*. Routledge.

Tye, Michael. 2009. *Consciousness Revisited*: *Materialism Without Phenomenal Concepts.* The MIT Press.

Unger, Roberto Mangbabiera. 2007. *The Self Awakened: Pragmatism Unbound.* Harvard University Press.

Urmson, J. O. and Jonathan Ree, eds. 2000. *The Concise Encyclopedia of Western Philosophy and Philosophers*. Routledge.

Velkley, Richard I. 2002. *Being After Rousseau: Philosophy and Culture in Question*. University of Chicago Press.

Vernant, Jean-Pierre. 2006. Janet Lloyd and Jeff Fort, trans. *Myth and Thought Among the Greeks*. Zone Books.

Vidal-Naquet, Pierre. 1999. in M. Detienne, Janet Lloyd, trans. *The Masters of Truth in Archaic Greece*. Zone Books.

Voegelin, Eric. 1990. Gerhart Neimeyer, trans. *Anamnesis*. University of Missouri.

Wachterhauser, Brice. 2002. In Robert J. Dostal, ed. *The Cambridge Companion to Gadamer*. Cambridge University Press.

Wallace, Alan B. 2012. *Dreaming Yourself Awake: Lucid Dreaming and Tibetan Dream Yoga for Insight and Transformation.* Shambala.

Watson, Richard A. 1999. In Robert Audi, general ed. *Cambridge Dictionary of Philosophy*, 2nd ed. Cambridge University Press.

Wegner, Daniel M. 2002. *The Illusion of Conscious Will*. The MIT Press.

Weiner, Norbert. 1948. *Cybernetics.* John Wiley and Sons.

———. 1950. *The Human Use of Human Beings: Cybernetics and Society*. Doubleday Anchor Books.

Wilson, E. O. 1998. *Consilience: The Unity of Knowledge*. Knopf.

Winn, Ralph B. 1960. *Dictionary of Existentialism.* Philosophical Library.

Wittgenstein, Ludwig. 1921. D. F. Pears and B. F. McGuinness, trans., *Tractatus Logico-Philosophicus.* Routledge.

———. 1958. G. E. M. Anscombe, trans. *Philosophical Investigations*, 3d ed., Prentice Hall, Inc.

———. 1969. Denis Paul and G. E. M. Anscombe, trans. *On Certainty.* Harper and Row.

Wood, David. 1989. *The Deconstruction of Time.* Northwestern University Press.

Yates, Frances A. 1966. *The Art of Memory.* The University of Chicago Press.

Yourgrau, Palle. 2005. *A World Without Time: The Forgotten Legacy of Gödel and Einstein.* Basic Books.

Zawidzki, Tadeusz. 2007. *Dennett.* Oneworld Publications.

Zeh, H. Deiter. 1999. *The Physical Basis of the Direction of Time.* Springer-Verlag.

INDEX